Convergence to Cosmicrobia

The Final Acceptance of Life as a Cosmic Phenomenon

Other World Scientific Titles by the Author

Comets and the Origin of Life
ISBN: 978-981-256-635-5

A Journey with Fred Hoyle
Second Edition
ISBN: 978-981-4436-12-0 (pbk)

The Search for Our Cosmic Ancestry
ISBN: 978-981-4616-96-6
ISBN: 978-981-4616-97-3 (pbk)

Where Did We Come From?: Life of an Astrobiologist
ISBN: 978-981-4641-39-5
ISBN: 978-981-4641-40-1 (pbk)

Vindication of Cosmic Biology:Tribute to Sir Fred Hoyle (1915–2001)
ISBN: 978-981-4675-25-3

Proofs that Life is Cosmic: Acceptance of a New Paradigm
ISBN: 978-981-323-310-2

Diseases from Outer Space — Our Cosmic Destiny
Second Edition
ISBN: 978-981-12-2212-2

Understanding the Origin and Global Spread of COVID-19
ISBN: 978-981-12-5907-4

Life Comes from Space: The Decisive Evidence
ISBN: 978-981-12-6625-6

CONVERGENCE TO COSMICROBIA
The Final Acceptance of Life as a Cosmic Phenomenon

Edited by

Chandra Wickramasinghe
University of Buckingham, UK

Rudolph Schild
Harvard and Smithsonian Center for Astrophysics, USA

J H (Cass) Forrington

NEW JERSEY • LONDON • SINGAPORE • BEIJING • SHANGHAI • HONG KONG • TAIPEI • CHENNAI

Published by

World Scientific Publishing Co. Pte. Ltd.
5 Toh Tuck Link, Singapore 596224
USA office: 27 Warren Street, Suite 401-402, Hackensack, NJ 07601
UK office: 57 Shelton Street, Covent Garden, London WC2H 9HE

Library of Congress Control Number: 2024943397

British Library Cataloguing-in-Publication Data
A catalogue record for this book is available from the British Library.

CONVERGENCE TO COSMICROBIA
The Final Acceptance of Life as a Cosmic Phenomenon

Copyright © 2025 by World Scientific Publishing Co. Pte. Ltd.

All rights reserved. This book, or parts thereof, may not be reproduced in any form or by any means, electronic or mechanical, including photocopying, recording or any information storage and retrieval system now known or to be invented, without written permission from the publisher.

For photocopying of material in this volume, please pay a copying fee through the Copyright Clearance Center, Inc., 222 Rosewood Drive, Danvers, MA 01923, USA. In this case permission to photocopy is not required from the publisher.

ISBN 978-981-98-0088-9 (hardcover)
ISBN 978-981-98-0089-6 (ebook for institutions)
ISBN 978-981-98-0090-2 (ebook for individuals)

For any available supplementary material, please visit
https://www.worldscientific.com/worldscibooks/10.1142/14042#t=suppl

Typeset by Stallion Press
Email: enquiries@stallionpress.com

Prologue

In 1989, Dr. David Latham at the Harvard-Smithsonian published his first detection of a "extra-solar planetary object candidate," from a detection made by the periodic radial motion variability method. The detection was soon confirmed by other astronomers and following observations by the NASA/KEPLER mission would soon yield thousands of additional discoveries, and a new science of extra-solar system planetary astronomy was launched. In the previous 2 decades, Chandra Wickramasinghe and Sir Fred Hoyle had already been pushing the boundaries of science by making the case that abundant research indicated that life could have originated in the distant reaches of our galaxy and some samples of it were reaching the Earth, as cosmic dust or enshrouded in meteorites, as fossilized specimens. This challenged the widespread but unproven belief that life was limited to within the solar system, and possibly unique to planet Earth.

In this book, Prof. Wickramamasinghe presents an update for new astronomical data obtained by a new generation of mountaintop telescopes and NASA space missions, as well as laboratory studies and geological discoveries.

Today, the James Webb Space Telescope is discovering spectroscopic and other markers and establishing that a galaxy at redshift $z = 12.6$ exhibits markers of life-related processes at the most distant reaches of the Universe known. Morre locally, evidence of complex structured chemical compounds (Buckey-Balls) is being found in clouds associated with star formation locally, indicating that this complex chemistry will be introduced into new generations of planetary systems from their birth.

Hence it is also time to ponder the perils our civilization faces from pollution of our biosphere by chance exposure to off-planet pollutants, even as we ponder the potentially beneficial advances possible from enhancements to the human genome with contact to a wider range of blessings.

Below: Bucky-Balls (NASA)

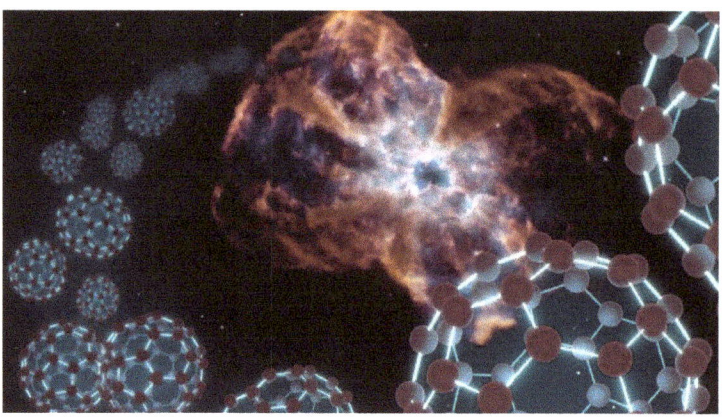

Content

Prologue   v

Introduction by Rudy Schild and Chandra Wickramasinghe   ix

Chapter 1	Cosmology and the Origins of Life	1
Chapter 2	Life Beyond the Limits of our Planetary System	15
Chapter 3	Quest for Life on Jupiter and its Moons	19
Chapter 4	Reluctance to Admit we are not Alone as an Intelligent Lifeform in the Cosmos	25
Chapter 5	The Second Copernican Revolution	37
Chapter 6	Search for Aliens, and UFO's	49
Chapter 7	A Note on a Biological Explanation for the ERE Phenomenon	59
Chapter 8	Cosmicrobia: A New Designation for the Theory of Cosmic Life	67
Chapter 9	Search for UFOs and Aliens: Modern Evidence and Ancient Traditions	75
Chapter 10	Life and the Universe: a Final Synthesis	87
Chapter 11	Standard Big-Bang Cosmology Faces Insurmountable Obstacles?	101

Epilogue   107

Bibliography   109

Introduction by Rudy Schild and Chandra Wickramasinghe:

A Second Copernican Revolution

Forty years ago, Fred Hoyle and Chandra Wickramasinghe first launched their challenge of what was essentially considered as the bedrock of Western science. They had begun to question the time-hallowed dogma of "the primordial soup on the Earth" as being the site and mode of origin of life. This was by no means a step they took lightly, being fully aware of the societal stigma that would inevitably follow to haunt them. The idea of life emerging spontaneously on Earth from inorganic material had formed the central core of Aristotelean philosophy dating back to the 3rd century BCE, a philosophy that had dominated Western thinking for over two millennia.

From the 1980's onwards, a veritable tide of new facts, driven by new space technologies and new telescopes, Hoyle and Wickramasinghe moved unerringly in the direction of challenging this time-hallowed point of view. Meticulously assessing the new facts to emerge from both astronomy and biology, and publishing such assessments in well over a hundred publications (many in the Journal *Nature*), these authors began to seriously challenge the prevailing ideas of an Earth-bound origin of life. It was argued that the first origin of life required a volume of space that far exceeded the scale of the solar system, the scale of the galaxy or even extending to much larger cosmological spaces. Once life has originated, however, its persistence through the mechanisms of "panspermia" seemed to be inevitable. It is precisely at this point that Hoyle and Wickramasinghe *predicted* the inextricable merger of astronomy and biology, eventually leading to the birth of the new discipline of astrobiology. ("The case for life as a cosmic phenomenon", F. Hoyle & N.C. Wickramasinghe, *Nature,* **322,** 509, 1986).

Although evidence in support of the idea of panspermia and cosmic life continued to grow from new developments in biology, geology and space science, societal disapproval showed little signs of abating. We should note here that a related paradigm that took centuries to overturn also had Aristotelean roots – the premise that the Earth was the centre of the solar system and the Universe. The Copernican revolution in astronomy that ultimately changed the status quo started with Galileo and Copernicus and concluded with Kepler and Newton, the whole process spanning the time period 1500-1700CE - nearly 200 years.

The first Copernican revolution displaced the Earth from its hallowed status of centrality in the Universe. The developments in astrobiology and cosmology over the past four decades define a second Copernican-style revolution which is now rapidly moving from the long-held idea of Earth-centred biology to the concept of life being a truly cosmic phenomenon.

In the papers constituting Vol.30 of the *Journal of Cosmology*, which are reprinted in this book, we find the strongest indications thus far that we are on the threshold of a major paradigm shift in Science. In fact, it would appear that discoveries made with new telescopes such as the James Webb Space Telescope have put the standard cosmological as well as biological theories on notice. All indications are that a major paradigm shift is just around the corner.

Cosmology and the Origins of Life

N. Chandra Wickramasinghe[1,2,3,4], Jayant V. Narlikar[5] and Gensuke Tokoro[4]

1. Buckingham Centre for Astrobiology, University of Buckingham, UK
2. Centre for Astrobiology, University of Ruhuna, Matara, Sri Lanka
3. National Institute of Fundamental Studies, Kandy, Sri Lanka
4. Institute for the Study of Panspermia and Astroeconomics, Gifu, Japan
5. Inter-University Centre for Astronomy and Astrophysics, Pune 411 007, India.

New evidence related to the origins of life in the cosmos combined with continuing progress in probing conditions of the early universe using the James Web Telescope suggest that long-held orthodox positions may be flawed. Only by objective evaluating the new facts and recognising the cultural forces at work can further progress be made **towards resolving perhaps the most important and fundamental questions in science.**

Keywords: Panspermia, interstellar dust, comets, cosmology

Introduction

In the modern technologically-driven world in which we live we tend to forget the cultural backdrop against which key scientific concepts are being rigidly maintained. These considerations do not apply however to well-established theories such as planetary dynamics and quantum physics, for example, that are rigorously based on prediction, experiment and verification. They apply to the much grander visions of cosmology and biology that lead to the boldest of assertions on how the universe and life within it arose. We shall point out that these latter pronouncements are by no means as secure as we are all too often made to believe.

The enlightenment in Europe in the 17th century heralded the beginning of the scientific method as well as the birth of scientific academies in Europe whose mission it was to put science on a firm rational and empirical basis. These developments served to stem the growth of superstition, magic and witchcraft that were rampant at the time. The advancement of empirical science that followed served us well for several centuries thereafter. However, over time, such benefits and advances began to act in a negative way by encouraging the rigid defence of scientific orthodoxies often against a tide of contrary evidence. Current attitudes to the origin of the universe as well as biology within it – the Big Bang theory and the theory of spontaneous generation of life on Earth - arguably fall into this latter category. Indeed, the ongoing insistence on defending scientific orthodoxies on these matters, even against a formidable tide of contrary evidence, has turned out to be no less repressive than the discarded superstitions in earlier times. **For instance, although all attempts to demonstrate spontaneous generation in the laboratory have led to failure for over half a century, strident assertions of its necessary operation against the most incredible odds continue to dominate the literature (6).**

Modern scientific ideas relating to the origin of life and the origin of the universe are directly traceable to Eurocentric philosophies that had developed mainly in the timespan between Aristotle (3rd century BCE) and St Thomas Aquinas (1225-1272CE). It should come as no surprise that St Thomas Aquinas accepted the entire Aristotlean corpus in so far as it related to Christian theology. The clashes with astronomical observations challenging geocentric cosmologies involving Galileo, Copernicus and Geodarno Bruno are of course well known. Perhaps less well known is the acceptance of the Aristotlean idea of the spontaneous generation of life – *fireflies emerging from mixtures of warm Earth and morning dew* – which forms the cornerstone of biology and persists in a modern form under the name of "abiogenesis".

Aristotle's idea of spontaneous generation of life posed a direct challenge to an earlier idea – panspermia – attributed to the pre-Socratic Greek philosopher Anaxoragas of Clazomenae who lived around 500BCE. Panspermia implies that the "seeds of life" are eternally present in the cosmos and takes root whenever and wherever the condition permit. Closely similar ideas prevailed in ancient India many centuries earlier for instance in the Vedas positing life to be an integral part of the structure of the universe.

Pasteur and life in the cosmos

The first serious attempts to re-examine spontaneous generation and to investigate panspermia from an experimental standpoint began with the French biologist Louis Pasteur in the early 1860's (1). Pasteur showed by means of laboratory experiments that what was already known for larger visible life forms - that life is always derived from pre-existing life of a similar kind. This casual chain of events – life from life - is true not only for life forms existing today but it is also true throughout the record of fossilised life on the Earth. The question that next arose already in the early 20th century is: when and where this connection cease to operate. We are then forced logically to conclude that the chain of "panspermic" connection continues "for ever" which in turn demands a cosmology that must also continue "for ever" in some form.

This connection has been discussed by several contemporary physicists. For instance, the German physicist Hermann von Helmholtz (2) wrote:

"It appears to me to be fully correct scientific procedure, if all our attempts fail to cause the production of organisms from non-living matter, to raise the question whether life has ever arisen, whether it is not as old as matter itself, and whether seeds have not been carried from one planet to another and developed everywhere where they have fallen on fertile soil..."

And in Britain Lord Kelvin (William Thomson) at about the same time declared "Dead matter cannot become living without coming under the influence of matter previously alive. This seems to me as sure a teaching of science as the law of gravitation......" (Ref.3)

Despite these enlightened responses that was followed by the championing of panspermia by Svante Arrhenius (4), a rigid orthodoxy advocating spontaneous generation prevailed well into the 20th century.

Problems with Spontaneous Generation

Fred Hoyle and one of us were perhaps the first to revive a serious interest in challenging spontaneous generation bringing forward a new case for panspermia. The first criticism of spontaneous generation that was voiced in the early 1980's (4,5) related to issues of probability of assembly of the crucial monomers of biology into a primitive living system. This was necessary to discuss because the chemical building blocks of life were being discovered to exist in vast quantity throughout the universe. The analogy that was made to a tornado blowing through a junk yard assembling an air plane is just one metaphor that came to be deployed to drive home the improbability of the transition from life molecules (monomers) to the simplest replicable lifeform.

Over the past few decades biologists have further unravelled the mind-blowing complexity of life at the molecular level and consequently laid bare its super-astronomical information content. Such a complexity is manifest for instance in the arrangements of amino acids in crucial enzymes, or nucleobases in DNA. The precise "information" contained in enzymes—the arrangements of amino acids into folded chains—is transmitted by way of the coded ordering of the four nucleotide bases (A,T,G,C) in DNA. In a hypothetical RNA world, that some biologists think may have predated the DNA-protein world, RNA is posited to serve a dual role as both enzyme and transmitter of genetic information. If a few such ribozymes are regarded as precursors to all life, one could attempt to make an estimate of the probability of the assembly of a simple ribozyme composed of 300 bases. This probability turns out to be 1 in 4^{300}, which is equivalent to 1 in 10^{180}, which can hardly be supposed to happen even once in the entire 13.9-billion-year history of the canonical Big Bang universe. And this is just for a single enzyme. In the simplest known bacterium *M. genitalium* with some 500 genes coding for enzymes the improbability escalates to a super-astronomical scale (6,7).

Geological and astrophysical evidence

Four decades ago the earliest evidence for microbial life in the geological record was thought to be in the form of cyanobacteria-like fossils dating back to 3.5 Ga ago. From the time of formation of a stable crust on the Earth 4.3 Ga ago following an episode of violent impacts with comets (the Hadean Epoch) there seemed to be available an 800 million years timespan during which the canonical Haldane-Oparin primordial soup and the spontaneous generation of life may have arguably developed. Very recent discoveries, however, have shown that this time interval has been effectively closed. Ancient rocks laid down 4.2 billion years ago belonging to a geological outcrop in the Jack Hills region of Western Australia have been found to contain micron-sized graphite spheres with an isotopic signature of biogenic carbon – fossil bacteria at a time when the collisions of the Earth with comets and asteroids were happening at a relentless pace (8). The requirement now, on the basis of orthodox thinking, is that an essentially instantaneous transformation of non-living organic matter to bacterial life took place, a proposition that strains credibility of Earth-bound abiogenesis to its utmost limit as we have already noted. A far more plausible proposition, in the light of the new evidence, is that fully-developed microorganisms arrived at the Earth via impacting comets, and these became carbonized and trapped within ancient rocks.

The crucial step from non-living organic molecules to primitive life-forms – bacteria and viruses – that can carry the entire range of possibilities for evolution of life could not, however, have happened on the Earth, on the surface of any planet, comet or asteroid nor indeed on any other restricted astrophysical setting. The evidence, in our view, points to such a transformation being linked to cosmology on the largest possible scale.

Since the early 1980's astronomical evidence has steadily accumulated that point inexorably to interstellar dust having a distinct biological provenance (4,5,9,10). Spectral features spanning the spectrum from the mid-infrared, visual and ultraviolet wavelengths have shown consistency with biological material in various states of degradation. A selection of the key astronomical data is displayed in Fig. 1 – points representing the data, and the curves the theoretical fits based on the omnipresence of bacteria, viruses, and their degradation products.

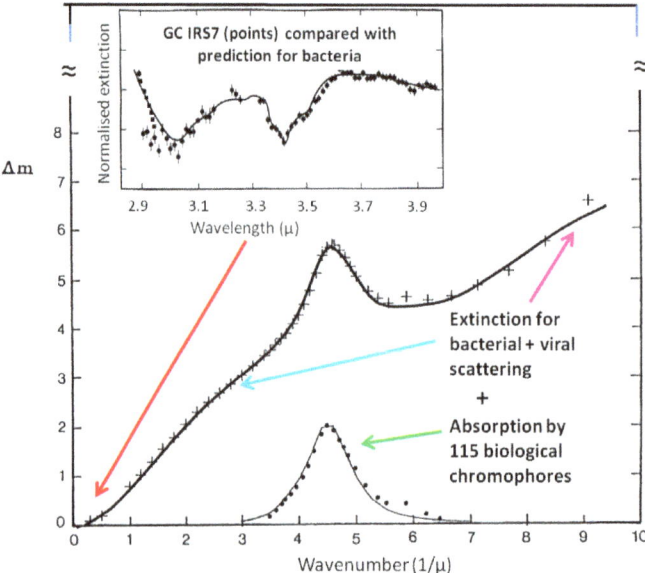

Fig. 1. *Upper Curve Main frame* The mean extinction curve of the galaxy (points) compared with the contribution of desiccated bacteria and nanobacteria.
Lower curve Main frame The residual extinction (points) compared with the normalized absorption coefficient of an ensemble of 115 biological aromatic molecules.
Inset: The first detailed observations of the Galactic centre infrared source GC-IRS7 (Allen & Wickramasinghe 1981) compared with earlier laboratory spectral data for dehydrated bacteria. (See citations in refs 9,10)

The first strong hint of a biological connection emerged in the absorption spectrum of galactic dust over a 10kpc pathlength from the Galactic Centre source GC-IRS7 shown in the inset of Fig.1. An absorption profile in the 2.9-4 micrometre wavelength range that was *predicted* for a bacterial component of interstellar dust was found to be present in the interstellar when the first observations (points in the inset) were subsequently made. The main panel shows a broad ultraviolet absorption feature of interstellar dust centred on the wavelength 2175A that was attributed to aromatic molecules in biology (see citations in 9,10). This feature attributable to biological dust shows up not only in our galaxy but in external galaxies as well.

In the context of our claims of a match between astronomical spectral data and an all-pervasive cosmic microbiology such as illustrated in Fig.1 and later in Fig.3, we stress that it is not the simple correspondence of absorption wavelengths in individual functional groups within organic molecules that is claimed here, but rather the integrated absorption/emission spectrum of an entire ensemble of organic functional groups as occurs in a bacterium of a virus. This is not a test used normally by chemists in laboratory spectroscopy, but in the present context remains a powerful argument in support of biology prevailing on a cosmic or cosmological scale.

Another set of astronomical data that has not been explained in nearly 100 years are the diffuse interstellar absorption bands in the optical spectra of stars. The strongest of these is centred at 4430A and has a half-width of ~ 30A. A possible solution to a 100-year old unsolved problem may also be connected with the behaviour of fragmentation products of biology existing under various states of excitation in their electronic configurations. A possible candidate in this category was originally proposed by F.M. Johnson in the form of magnesium tetrabenzo porphyrin (11). More recently a set of other infrared absorption bands in interstellar dust has been found to exist in our galaxy as well as in external galaxies. These are attributed to "polyaromatic hydrocarbon", PAH's but this designation does not explain their origin. A large fraction of the "PAH's" and other organic molecules discovered in the galaxy as well as in external galaxies in our view could represent biological material in various stages of degradation.

The total mass of material tied up in the form of molecules that could have a biological connection in the galaxy (and in the wider cosmos) amounts to possibly a third of all the available carbon – a fraction of a percent of mass of the entire galaxy (12). The question then arises as to whether these biologically relevant molecules so widely present in the cosmos represent steps towards life – prebiotic evolution – or whether they are the products of biological degradation – the detritus of life. The overwhelming bulk of the organic material we find on Earth is unequivocally the result of the decay of biology. So, the question we need to ask is this: Why is it not the same for the organics in space?

Critics who are culturally opposed to think of life as a cosmic phenomenon regard panspermia an "extraordinary hypothesis" and it is stated that extraordinary evidence would be needed to defend such an extraordinary idea. We claim that, on the contrary, confining life to Earth could be regarded as a far more extraordinary assertion, so it is the defence of this latter point of view that must require extraordinary evidence. And such evidence is non-existent, or at best illusory.

The overriding rationale for interstellar dust grains, or a significant fraction thereof, being connected with biology stems from the argument that life itself could only have arisen in a cosmological setting – requiring a volume of space that transcends enormously the miniscule scale of our planet. We then proceed to argue that a cosmologically derived legacy of life along with its full evolutionary potential (contained within the genomes of bacteria and viruses) were introduced via comets onto habitable planets like the Earth in our Milky Way system and beyond. Microbial life thereafter is amplified and recycled between billions of planetary abodes, of which our solar system is just one. Microbial material on this picture must escape continuously into the interstellar medium from comets and planetary systems as indicated in the feedback loop in Fig. 2.

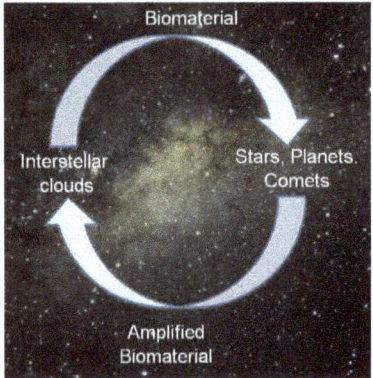

Fig.2 Bacteria and viruses expelled from a planetary system are amplified in the warm radioactively heated interiors of comets and thrown back into interstellar space, where a fraction breaks up into molecular fragments that are observed, but a non-negligible minute fraction remains viable.

Comets

The first infrared spectrum of a comet, Comet P/Halley observed in March 1986, showed consistency with bacterial dust emanating from an eruption of the comet (12). More recent studies of other comets have yielded generally similar results. Recently the European Space Agency's Rosetta Mission to comet 67P/C-G has provided the most detailed observations that satisfy all the consistency checks for biology and the theory of cometary panspermia. Fig. 3 shows the close consistency between the surface reflectivity of the comet at infrared wavelengths compared with properties of a desiccated bacterial sample (13).

Chapter 1 — Cosmology and the Origins of Life

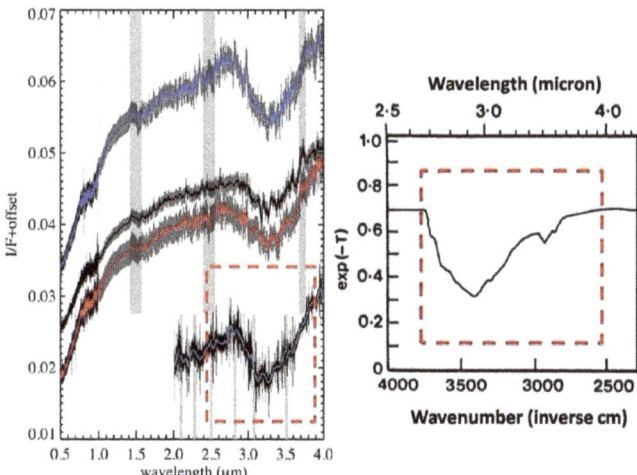

Fig. 3. The surface reflectivity spectra of comet 67P/C-G (left panel) compared with the transmittance curve measured for *E-coli* (right panel).

It seems ironical that the Rosetta Mission to Comet 67P/Churyumov-Gerasimenko that carried the lander Philae touching down with experiments such as produced Fig.4 did not include a life detection experiment. Some of us, who served on one of the Rosetta mission science teams, had proposed such an experiment similar to the 1976 Mars Viking labelled release detection experiment (14); but as expected this proposal was not included. In the event all we have is tantalising indirect evidence of life as for instance seen in Fig.3 that the critic can choose to ignore as being coincidence. The trouble comes when the number of such "coincidences" escalates to a point where such an assertion begins to look more and more difficult to defend.

More recently the discovery of a giant comet (C/2014 UN271) some 100km in diameter at a distance of 29AU in October 2014 and the later discovery in September 2021 of a dramatic brightening episode offers a further opportunity for verifying the predictions of fermentation processes in a "biological" comet (15) The eruptions of the comet at a heliocentric distance of 20AU (two thirds of the distance from the sun to Neptune) can only be plausibly explained as due to high pressure venting of the products of microbial metabolism in radioactively heated subsurface lakes.

Carbonaceous meteorites (residues of comets) and other bodies in the solar system have also come under close scrutiny over the past few decades. Space Missions combined with laboratory investigations have provided clear evidence for liquid water and indigenous extraterrestrial organics and biomolecules in carbonaceous chondrites as well as in low density asteroids (Hoover et al, 2022; ref 16). Recent discoveries of biomolecules including amino acids and nucleobases (purine and pyrimidine) in some carbonaceous asteroids and meteorites (17) have been hailed as supportive evidence for a biological connection, although in a limited and in our view flawed context of supplying merely the components of a primordial soup on Earth. Scanning Electron Microscope studies over the past few decades have provided clear evidence of indigenous microfossils in diverse groups of carbonaceous meteorites but they tend to be ignored or dismissed. Thus, the long-held culturally sanctioned Aristotelean belief that terrestrial life must necessarily start *de novo* remains hard to shake off (4,7).

Stratospheric evidence

If comets are the repositories of cosmic life the question arises as to whether this proposition is open to test and verification at the present time. The Earth's orbit continually crosses streams of cometary debris so is it possible to detect bacteria in our neighbourhood, perhaps in the stratosphere? To answer this question one of us (JVN) approached the Indian Space Research Organisation (ISRO) in 2000 seeking their collaboration to make the first carefully controlled recovery of microbial structures (bacteria and putative viruses) from a height of 41km in the stratosphere, presumably falling in from space and of extraterrestrial/cometary origin.

The startling conclusion from this sampling experiment was that positive detections of in-falling microbiota collected in a measured volume of the stratosphere at 41km led to an estimate of an in-fall rate over the whole Earth of 0.3-3 tonnes of microbes per day. This converts to some 20-200 million bacteria per square metre arriving from space every single day (18,19). Between 2001 and the present day this infall rate of microbiota would appear to have been amply confirmed, although not still widely admitted.

The results from the first ISRO-sponsored balloon flight in 2001 (see Shivaji et al. [20]) included a further significant finding. The bacteria isolated by CCMB (Centre for Cellular and Molecular Biology, India) were uncannily resistant to ultra violet radiation. This indicated that these microorganisms evolved in the presence of UV radiation, thus making it very unlikely that they came from the surface of the Earth with only a short residence time (without replication) in the stratosphere.

Following the first 2001 ISRO balloon launch a second stratospheric sampling flight was carried out in 2005 (20,21) when 12 species of bacteria were found at a height of 41km in the stratosphere. Of these three were entirely new in the sense that they had never been identified on the Earth. They were named after Fred Hoyle, Aryabhata (the 5th century Indian astronomer) and ISRO (21). Currently an ISRO-based team is in the process of using nanotechnology to devise ways of isotope analysis to distinguish unequivocally between terrestrial and extraterrestrial microorganisms.

The canonical cosmological context

We have consistently argued over many years that in order to understand the origin of the super-astronomically vast and complex informational system as we find in biology it is imperative to go to the "biggest" available system in which such information can be generated. That system is unquestionably the entire universe, and so biology and cosmology must come to be understood as being inextricably linked. The cosmological backdrop against which this link has to be understood has evolved over a nearly a century. In 1929 Edwin Hubble discovered the relationship between the distances of faraway galaxies and the redshift of their spectral lines. The latter was interpreted as Doppler shift due to recession with a connection thus established between velocity of recession – speed at which galaxies were rushing away from one another - and their absolute magnitude or intrinsic luminosity. This led to the concept of an expanding universe, with "Hubble's law" defining the rate at which the universe appears to be flying apart from an initial origin as a "point". Reversing the speeds of expansion as indicated by Hubble's law soon led to an estimated age of the universe of some 13.7 billion years and to the concept of "Big Bang Cosmology" – everything we find in the universe starting off as a point. Such a mathematical point containing all the energy and all the information for physics as well as biology remains a concept of origination that by its very essence cannot be further explored. However, in one form or other this is the model of the universe that has come down to be regarded as the scientific orthodoxy which everyone is supposed to accept in the present day. It will be disingenuous to deny that cultural considerations have played a decisive role in controlling the scientific narrative. To sum up, the prevailing orthodoxies in both cosmology and biogenesis continue a strictly Aristotelean tradition, which incidentally accords well with a Judeo-Christian world view – "God said let there be light and…."

The discovery of the *cosmic microwave background* radiation (CMB) by Penzias and Wilson (22) came to be regarded as the transformative evidence supportive of the standard Big Bang theory of the origin of the universe - the afterglow of light and radiation left over from the infinitely hot Big Bang cooled down to a temperature of 3.5K. It is this evidence that was used to challenge the rival steady-state cosmology advocated by Hoyle and one of us (JVN) in subsequent decades. Alternative mechanisms for explaining the microwave background data were never popular, and the Big-Bang cosmological model, sanctioned by a strong cultural tradition, became firmly rooted in astronomy. The autobiographical reviews by one of us (JVN) will serve to outline how scientific prejudice operates [23-27]. The 'bang wagon' effect has led to a 'confirmed belief' in big bang cosmology. In this context we also commend in particular an article by Fred Hoyle [26] giving a 'mathematical' analysis of how belief in a paradigm grows despite lack of real evidence.

The progression of energy/matter from the postulated Big Bang event to atoms, stars and galaxies has been fertile ground for cosmologists over many decades. The first phase known as inflation leads to a rapid succession of doublings the formation of light elements a quiescent phase known as the "dark ages", re-ionisation, and the formation of the first stars and galaxies. The last of these steps was not conceived of as happening prior to some 400 million years following the "Big Bang" event.

Alternative cosmological models

It is not widely recognised in the scientific world that ideas relating to an infinite age of the universe and models involving an infinite sequence of creation and destruction episodes have a distinct Indian provenance.

According to Carl Sagan:

"The Hindu religion is the only one of the world's great faiths dedicated to the idea that the Cosmos itself undergoes an immense, indeed an infinite, number of deaths and rebirths. It is the only religion in which time scales correspond to those of modern scientific cosmology. Its cycles run from our ordinary day and night to a day and night of Brahma, 8.64 billion years long, longer than the age of the Earth or the Sun and about half the time since the Big Bang. And there are much longer timescales still......"

(Cosmos: The Story of Cosmic Evolution, Science and Civilisation. 1983)

Jain and Buddhist traditions (500 BCE) also follow the same thought with cyclical universe and were most probably continued from the earlier Hindu texts.

In this context it is worth reiterating that the currently favoured Big-Bang theory of the Universe with an age of 13.8 billion years is by no means absolutely proved. The very recent discovery of a galaxy designated GN-z11 located at a distance of 13.4 billion light years (implying its formation just 420 million years after the posited Big Bang origin of the Universe) poses serious problems for the current consensus view of cosmology (28). Similar problems for the Big Bang cosmological model have been discussed over a period of some 3 decades by small group of dissenters (29).

Recently Nobel Laureate Roger Penrose has come in among the select band of dissenters from the standard view of a unique Big Bang origin of the Universe 13.8 billion years ago (30,31). In a theory called the "conformal cyclic cosmology" Penrose postulates that the universe undergoes an infinite number of cycles in which the Big Bang event 13.8 billion years ago is the most recent cycle of which we are a part. The difference between the Penrose models and those of Hoyle and Narlikar involving quasi-steady-state cosmologies do not appear to be vast (24).

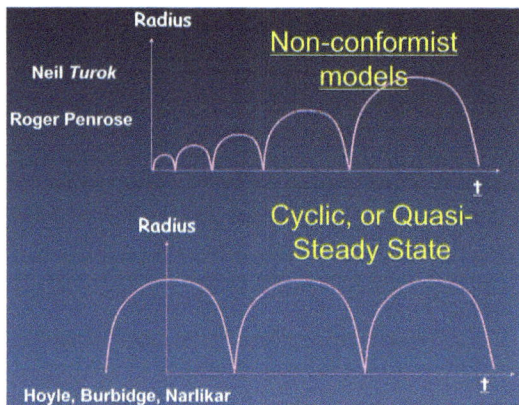

Fig.4. Schematic depiction of non-conformist models of the Universe
Both theories posit cyclic models that are in general coincidentally consonant with ancient Vedic-Hindu ideas.

The James Webb Space Telescope that came into operation this year was designed to look farther back in space and time than any other telescope, so it is not surprising that it may have detected the most distant galaxy in the cosmos (28). The object known by the designation CEERS-93316, a galaxy – if indeed it is confirmed as a galaxy – will be some 35 billion light-years away. We have now gone back in time from 400 million years after the Big Bang in the case of GN z-11 to a staggering 235 million years proximity to the Big Bang. This moment is only about 135 million years after the first stars are thought to have been born.

Fig. 5. CEERS-93316 presumed to be the most distant galaxy discovered thus far, implying it is now at a redshift of z=16.7 implying it is now at a distance of ~ 35 billion light-years from Earth (courtesy NASA) (28).

One might wonder how such a distance is plausible. The universe is only 13.8 billion years old, how can anything be farther away than that? The answer is that the universe has expanded greatly since the light first left the galaxy about 13.6 billion years ago, so that the "proper distance" to CEERS-93316 now is in fact 35 billion light-years. This new data already poses severe problems for standard Big-Bang models of the Universe. But there would surely be more surprises in store – even older galaxies where none is expected, and hopefully spectroscopic data clearly pointing to life at the very dawn of time.

References

1. Pasteur, Louis, 1857. Comptes rendus de l'Académie des Sciences, 45, 913.

2. von Helmholtz, in W. Thomson & P.G. Tait (eds).. 1874 *Handbuch de Theoretische Physik*, Vo.1. Part 2., Brancsheig.

3. Thompson, W. (Lord Kelvin), 1871. British Association for the Advancement of Science, Presidential address.

4. Arrhenius, S., 1908. Worlds in the Making, Harper, Lond.

5. Hoyle, F. and Wickramasinghe, N.C., 2000. Astronomical Origins of Life: Steps towards Panspermia (Kluwer Academic Press)

6. Wickramasinghe, N.C., 2013. Simulation of Earth-based theory with negative results, BioScience, 63(2), 141

7. Wickramasinghe, N.C., 2010 Astrobiological Case for our Cosmic Ancestry. *International Journal of Astrobiology*, 9(2), 119

8. Bell, E.A., Boehnke, P., Harrison, T. et al, 2015. Potentially biogenic carbon preserved in a 4.1 billion-year-old zircon, PNAS, 112 (47) 14518-14521

9. Wickramasinghe, N.C., Wickramasinghe, D.T., Tout, C.A., 2019. Cosmic Biology in Perspective, Astrophysics and Space Science, 364, 205.

10. Steele EJ, Al-Mufti S, Augustyn KK, Chandrajith R, Coghlan JP, Coulson SG, Ghosh S, Gillman M. et al 2018 "Cause of Cambrian Explosion: Terrestrial or Cosmic?" Prog. Biophys. Mol. Biol. 136: 3-23, https://doi.org/10.1016/j.pbiomolbio.2018.03.004

11. Johnson, F.M., 2006. Diffuse interstellar bands: a comprehensive laboratory study, Spectrochim. Acta A. Molecular Biology Spectroscopy 65(5) 1154-1179

12. Wickramasinghe, D.T. and Allen, D.A., 1986. Discovery of organic grains in Comet Halley, *Nature*, 323, 44-46

13. Capaccione F, Coradini A, Filacchione G., et al 2015, The organic-rich surface of comet 67P/Churyumov-Gerasimenko as seen by VIRTIS/Rosetta, *Science* 347 (6220).

14. Levin, G.V. and Straat, P.A., 2016. The case for extant life on Mars and its possible detection in the Viking Labelled Release Experiment, Astrobiology, 16, 798-810

15. Wickramasinghe, N.C. 2022. Giant comet C/2014 UN271 (Bernardinelli-Bernstein) provides new evidence for cometary panspermia, *International Journal of Astronomy and Astrophysics,* 12, 1-6

16. Hoover, R.B. et al, 2022. ENAA and SEM Investigations of Carbonaceous Meteorites: Implications to the Distribution of Life and Biospheres, Academia Letters

17. Oba, Y., Takano, Y., Furukawa, Y. et al, 2022. Identifying the wide diversity of extraterrestrial purine and pyrimidine nucleobases in carbonaceous meteorites, Nature.Comms, 13, 2008

18. Harris, M.J., Wickramasinghe, N.C., Lloyd, D., Narlikar, J.V., et al (2001) The detection of living cells in stratospheric samples. Proc SPIE 4495, 192–198.

19. Lloyd, D., Wickramasinghe,N.C., Harris, M.J., Narlikar, J.V.et al, (2002. Possible detection of extraterrestrial life in stratospheric samples. Proceedings of the International Conference entitled "Multicolour Universe" eds R.K. Manchanda and B. Paul (Tata Institute of Fundamental Research, Mumbai, India, pages 367ff

20. Shivaji, S., Chaturvedi, P., Begum, Z. et al, 2009. *Janibacter hoylei sp. nov.*, *Bacillus isronensis sp. nov.* and *Bacillus aryabhattai* sp. nov., isolated from cryotubes used for collecting air from the upper atmosphere. *Int. J Systematics Evol. Microbiol*, 59, 2977–2986

21. Shivaji, S., Chaturvedi, P., Suresh, K. et al, 2006. Bacillus aerius sp. nov., Bacillus aerophilus sp. nov., Bacillus stratosphericus sp. nov. and Bacillusaltitudinis sp. nov., isolated from cryogenic tubes used for collecting air samples from high altitudes, *International Journal of Systematic and Evolutionary Microbiology*, **56**, 1465

22. Penzias, A.A.; R. W. Wilson, 1965. A Measurement Of Excess Antenna Temperature At 4080 Mc/s. Astrophysical Journal Letters. **142**: 419–421.

23. Narlikar, J.V., 2015. Trials and tribulations of playing the devil's advocate, *Research in Astronomy and Astrophysics*, **15**, 1, 1

24. Narlikar, J.V. ,2018. The evolution of modern cosmology as seen through a personalwalk across six decades, *The European Physical Journal H*, **43**, 1, 43

25.Narlikar, J.V., 2021. Three pathbreaking papers of 1966 revisited: their relevance to certain aspects of cosmological creation today, *The European Physical Journal H*,**46**:21

26. Hoyle, F., 1994. IUCAA-IAGRG Silver Jubilee Conference Proceedings:Opening Address; Editor P.S. Joshi

27. Hoyle, F., Burbidge, G. and Narlikar, J.V., 2008. A Different Approach to Cosmology: From a Static Universe through the Big Bang towards Reality (Cambridge University Press)

28. Brammer, P.A. and van Dokkum, P.G., 2016. A Remarkably Luminous Galaxy at z=11.1 Measured with Hubble Space Telescope Grism Spectroscopy, The Astrophysical Journal, 819(2), article id. 129

29. Narlikar, J.V., Burbidge, G., and Vishwakarma, R.G., 2007 Cosmology and Cosmogony in a Cyclic Universe, Astrophys. Astron., **28**, 67–99

30. Penrose Roger, 2006. "Before the Big Bang: An Outrageous New Perspective and its Implications for Particle Physics". Proceedings of the EPAC 2006, Edinburgh, Scotland: 2759–2762.

31. Gurzadyan, VG and Penrose, R, 2013. "On CCC-predicted concentric low-variance circles in the CMB sky". Eur. Phys. J. Plus. **128** (2): 22. arXiv:1302.5162

Life beyond the limits of our planetary system

N. Chandra Wickramasinghe[1,2,3,4]

1. Buckingham Centre for Astrobiology, University of Buckingham, UK
2. Centre for Astrobiology, University of Ruhuna, Matara, Sri Lanka
3. National Institute of Fundamental Studies, Kandy, Sri Lanka
4. Institute for the Study of Panspermia and Astroeconomics, Gifu, Japan

Summary

Evidence for the widespread distribution of biologically relevant molecules widely throughout the Galaxy and beyond has been in existence for many decades. The recent discovery of a nucleobase uracil adds to an already impressive body of evidence that supports a cosmic origin of the complex building blocks of life.

Keywords: Carbonaceous asteroids, organic molecules in space, origin of life, panspermia

The distance from Earth of the much publicised carbonaceous asteroid 162173 Ryugu (1999 JU3) at 2-3 AU in which biologically related organic monomers have been discovered (Oba et al, 2023) (1) pales into insignificance compared to the astronomical distances at which similar organic molecules have been discovered in the past. As early as the 1920's the discovery of the "diffuse interstellar bands" DIB's in the optical spectra of reddened stars pointed to organic molecules associated with interstellar dust extending over tens of kiloparsecs (Wickramasinghe, N.C., 1967(2); Herbig, G.H. 1995 (3)). This is shown in Fig.1 the absorption band at 4430A being the strongest and most persistent feature, with other strong features located at the wavelengths 5780, 5797, 6284 and 6614A.

In 1967 F.M. Johnson pointed out to the consternation of the astronomical community that a molecule related to chlorophyll, magnesium tetrabenzoporphirine ($MgC_{46}H_{30}N_6$) had an absorption spectrum that was fitted the DIB's to a remarkable degree of precision (Johnson, 1967 (4); Johnson et al, 1973(5)). Whilst this identification remains in dispute, other authors have considered the possibility of interstellar extinction arising from polyaromatic hydrocarbons (Donn, 1968 (6)), a possibility that appears to have been vindicated with the discovery of a broad 3.3 micrometre emission feature in a number of reflection nebulae (4. Sellgren et al, 1983)(7).

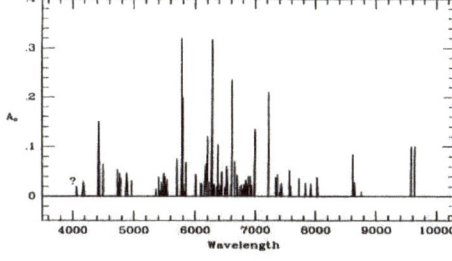 

Fig. 1 The diffuse interstellar bands *(L)* attributed by F.M. Johnson to magnesium tetrabenzoporphirine ($MgC_{46}H_{30}N_6$) *(R)*

From 1984 onwards the detection of discrete interstellar emission features in the infrared particularly those at wavelengths 3.3, 6.2, 7.7, 8.6, and 11.3 μm that have been generally attributed aromatic features such as exist in the combustion of biological products. These are found in both galactic and extragalactic sources has led to the acceptance of polyaromatic hydrocarbon molecules having a widespread cosmic prevalence (Wilner et al, 1977(8); Wilner, 1984(9)). One particular class of source that show these bands are planetary nebulae of which the source NGC2027 is a typical example at a distance close to 1 kpc. The IR spectrum of this source is shown in Fig. 2.

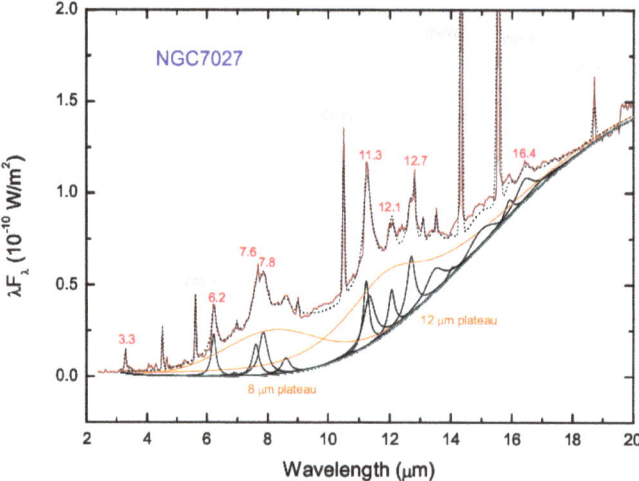

Fig.2 Positions of the PAH bands in the planetary Nebula NGC7027.

Mid-infrared spectra of HII regions within star-forming galaxies M83 and M33 at distances of the order of Mpc. However, it should be noted that any possible biological origin for these molecules in our galaxy and beyond still appears to be vigorously resisted for what the author believes are mainly cultural reasons.

As far back as 1974 present author suggested the identification of organic polymers in interstellar dust and comets typified by polyformaldeyde (Wickramasinghe, 1974 (10); Cooke and Wickramasinghe, 1977 (11); Vanysek and Wickramasinghe, 1977(12)). Whilst these ideas received some support from the detection formaldehyde in the gas coma of Halley's comet (Huebner, 1987 (13); Mitchel et al, 1987 (14), Hoyle and Wickramasinghe (1991) (15) next began to explore the possibility of biological related polymers including polysaccharides forming in the mass flows from stars.

The possibility of biological molecules – indeed entire biological structures such as freeze-dried bacteria existing on a galactic scale – followed from infrared spectroscopy when the absorption by laboratory systems were compared with new observations of astronomical sources at infrared wavelengths. Laboratory spectra of desiccated bacteria by Hoyle et al, 1982 (16) compared with the infrared spectrum of the Galactic Centre source GC-IRS7 obtained by Dayal Wickramasinghe and David Allen showed a fit that caused a sensation amongst astronomers as well as biologists (16).

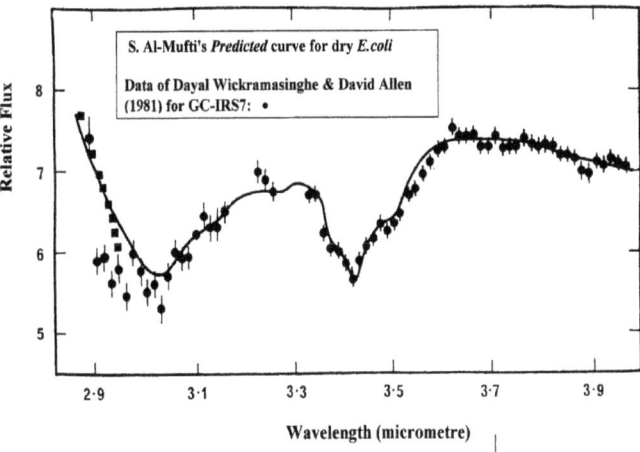

Fig.2. Astronomical data points compared with laboratory spectrum of dessicated bacteria under space-simulated conditions.

This match of the laboratory spectrum to astronomical data points shown in Fig 2 shows clearly that life-related biomolecules are present in cosmic dust over a distance scale of some 10kpc. A similar result obtained a few years later for the dust tail of Comet Halley during its 1986 perihelion (17) showed clearly that comets originating in the Oort cloud at a distance of 10's of thousands of AU from the sun also carried biologically related material. This is already in great excess of the distance of Ryugu (2 AU) at which material that contained biologically related molecules was recently recovered and analysed (1). The more radical position that has to be admitted is that the informationally rich components of life in the form of microscopic biological entities – bacteria and viruses - are present on a scale that transcends the size of planetary systems, star systems, even entire galaxies. It is high time we overcome cultural prejudice and accept that life is a truly cosmic phenomenon. Only then can new data as they become available be judged objectively for what they really mean.

References

1. Obo, Toga, Takano et al, 2023. Uracil in the carbonaceous asteroid 162173 Ryugu, Nature Comms., 14, 1292
2. Wickramasinghe, N.C, 1967, Interstellar Grains (Chapman & Hall, London)
3. Herbig, G.H., 1995. The diffuse interstellar bands, Ann.Rev.Astron.Astrophys., 33, 49-73
4. Johnson, F.M..1967. In J.M. Greenberg and T.P. Roark (eds) Interstellar Grains , NASP-140

5. Johnson, F.M., 1971, Annals N.Y. Acad. Scie., 194, 3
6. Donn, B., 1968. Astrophys.J.Lett., 152, L129
7. Selgren, K. et al, 1983. Astrophys.J., 271, L13
8. Willner, S.P. et al, 1979. Astrophys.J., 229, L65
9. Willner, S.P. 1984. In M.F. Kessler and J.P.Phillips (ed) Galactic and Extragalactic IR spectroscopy
10. Wickramasinghe, N.C., 1974. Nature, 252, 462
11. Cooke, A. and Wickramasinghe, N.C., 1977. Astrophys.Sp.Sci., 50, 43
12. Vanysek, V. and Wickramasinghe, N.C., 1975. Astrophys.Sp.Sci., 33, L19
13. Huebner, W.F., 1987. Science, 237, 628
14. Mitchel, D.L. et al, 1987, Science, 237, 626
15. Hoyle, F., Wickramasinghe, N.C.,1999 The theory of interstellar grain (Kluwer Dordrect)
16. Hoyle, F., Wickramasinghe, N.C., Olavesen, A.H., Al-Mufti, S., and Wickramasinghe, D.T.,1982. Astrophys.Sp.Sci., 83, 405
17. Wickramasinghe, D.T. and Allen, D.A., 1986. Discovery of organic grains in Comet Halley, Nature, 323, 44

Quest for life on Jupiter and its moons

N. Chandra Wickramasinghe[1,2,3,4] and Gensuke Tokoro[4]

1. Buckingham Centre for Astrobiology, University of Buckingham, UK
2. Centre for Astrobiology, University of Ruhuna, Matara, Sri Lanka
3. National Institute of Fundamental Studies, Kandy, Sri Lanka
4. Institute for the Study of Panspermia and Astroeconomics, Gifu, Japan

The final confirmation of the existence of multicellular life in aqueous habitats on the moons of Jupiter, will be a game changer for the societal approval and acceptance of panspermia which has been long overdue.

Keywords: Panspermia, comets, life in moons of Jupiter

On April 14, 2023 when the Sun moved into Aries marking a new astronomical year (a tradition observed in Sri Lanka and India), ESA's Jupiter Icy Moons Explorer, JUICE, was launched to make detailed observations of the giant planet Jupiter and its three large ocean-bearing moons – Ganymede, Callisto and Europa – to search for evidence of extraterrestrial life. The Juice spacecraft is due to take a leisurely 8 years to reach its Jovian destinations where indirect evidence for bio-friendly habitats have been known to exist for some time. NASA's Pioneer and Voyager probes made the same journey in 1970's and had already yielded a wealth of images and data that were arguably fully consistent with the presence of microbial life.

The theory of cometary Panspermia posits that the complex informational units of life in the form of genes (DNA, RNA), that can be assembled into life forms subject to constraints of natural selection, are distributed throughout the galaxy and beyond (Hoyle and Wickramasinghe, (1981a(1); 1981b(2); 1983(3); 1985(4); 1986(5)) . Bacteria and viruses that can serve as genetic and evolutionary vectors for such a process are, according to this model, incubated and nurtured in the interiors of comets and icy bolides and asteroids. Within such a cosmic evolutionary scheme that bypasses an impossible bottle-neck of spontaneous generation in any single planetary site, it will be expected that self-similar lifeforms develop from genetic units (bacteria and viruses) that are cosmically widespread. The similarity of microbiota as well as more complex aquatic lifeforms that are found on the Earth will most probably be replicated in other "ocean bearing" domains on other planets and moons in our solar system and beyond. Impactors containing the same suite of genes that are responsible for the emergence of such life will be distributed widely throughout the solar system and beyond by processes discussed in detail by Wallis and Wickramasinghe (2004 (6)).

Convergence to Cosmicrobia: The Final Acceptance of Life as a Cosmic Phenomenon

Fig. 1 Covers of scientific monographs by Hoyle and Wickramasinghe developing the theory of cometary panspermia in 1981, 1985 and 1996 where explicit reference to evidence of life in Jupiter and its moons is discussed.

If one takes account of the wide range of ambient conditions that support microbes as well as simple multicellular organisms in multiple locations on the Earth, the prospects for life on the icy moons of the planets Jupiter and Saturn cannot be ignored. The relevant necessary conditions include the existence of long-lived radioactive heat sources in these bodies that could maintain subsurface oceans over billions of years (Hoover et al, 2022 (8)). These would be similar to conditions that prevail at great depths in the Earth's crust, in the Antarctic ice shelves and in the high temperature hydrothermal vents in the ocean floor, where both single-celled bacteria and relatively simple multicellular life forms are known to abound. The limits to multicellular life will only be constrained by the arrival of appropriate genetic programs (DNA) combined with the supply of nutrients and hospitable ambient conditions.

Comets are well known to exhibit emergence of jets and eruptions at heliocentric distances that are too large to be explained by solar heating (Fig.2). The prodigious output of dust and gas observed in the Rosetta Mission comet 67P/CG, comet Hale-Bopp at a heliocentric distance of 6.5 AU and recently the eruptions of the giant comet C/2014 UN271 at an even greater distance of 29AU (9) clearly supports the presence of subsurface biological activity. Likewise, NASA's New Horizons mission in its flyby study of Pluto and its moons in 2015 provided similar evidence of jets and eruptions in relation to the dwarf planet Pluto, again confirming that radiogenic heating is relevant in 1000km-sized icy bodies (trans-Neptunian objects TNO's not unlike giant comets) throughout the outer regions of the solar system (6,7,8,10). Similarly, Triton, often regarded as a "twin" of Pluto, is also similarly active and displays unambiguous evidence of restructuring and mobility on timescales well under millions of years, as can be inferred from the lack of evidence for meteorite impact craters. All the available data points to sporadic high-pressure release of material from subsurface domains that have the ability to support microbial life.

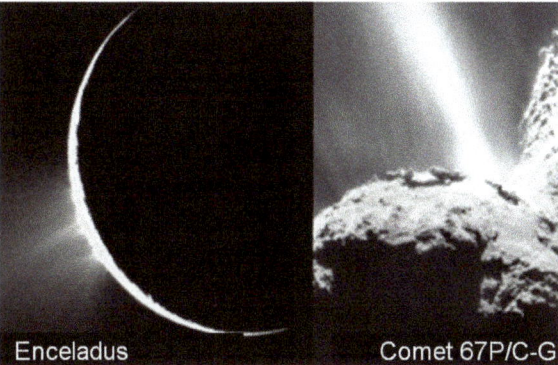

Fig. 2. Plumes of water and methane issuing from fissures in overlying ice. Left: Cassini image of the South pole of Enceladus; Rosetta image of jet of water and organics in comet 67P/C-G.

Viable habitats associated with Jupiter and Saturn have been under discussion for over 4 decades by Hoyle and the present writer (1). In our 1981 book, "Space Travellers – The Bringers of Life" (1), we discussed available data on dust in the clouds of Jupiter and concluded a consistency with bacteria and bacterial spores and an ongoing aerobiology operating in the Jovian atmosphere. We also considered at this time the hypothesis that the rings of Saturn contain bacterial dust expelled from its two icy satellites S13 and S14 that have retrograde orbits about the planet.

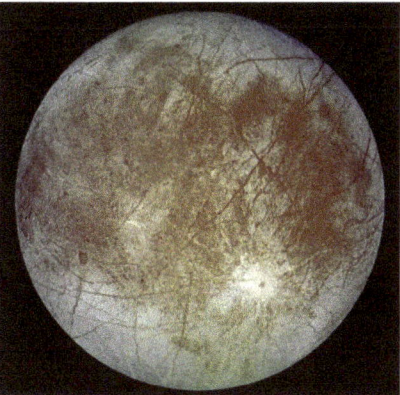

Fig. 3. Multi-cracked icy surface of Europa delineated by orange-coloured pigment

In 1996, after NASA's Galileo spacecraft had examined the surface of Jupiter's moon Europa and discovered a multi-cracked icy outer layer demarcated by orange pigments, we began to argue strongly for evidence of subsurface biology (Hoyle and Wickramasinghe (11)). This is still the most reasonable explanation for all the evidence on this Jovian moon available to date. It is remarkable that the orange coloration of the cracks ESA's is consistent with biological pigments, and evidence of radioactively and/or tidally heated subsurface oceans also remains strong. At the present time the case of a complex microbiota and/or multicellular life (small sea animals) existing on Europa cannot be stronger, and so the recent ESA/NASA missions to this Jovian moon cannot be more timely.

Fig.4. Author of *2001-A Space Odyssey*, Sir Arthur C. Clarke and Chandra Wickramasinghe discussing life on Europa (intelligent dolphins!) in the seas of Europa in 1981

We conclude by noting that in the book entitled "Greetings, carbon-based bipeds" by Science Fiction writer and futurist Arthur C. Clarke several bold predictions of the future are made, some of which have already come to true. Amongst others that still lie in waiting is his forecast that in 2061 in which he states:

"2061. The return of Halley's comet; first landing by humans. The sensational discovery of both dormant and active life-forms vindicates Hoyle and Wickramasinghe's century-old hypothesis that life is omnipresent throughout space....."

We would boldly predict that this realisation may well be achieved with JUICE a full 3 decades ahead of Arthur's forecast.

References

(1) Hoyle, F. and Wickramasinghe, N.C., 1981a. Comets - a vehicle for panspermia, in Ponnamperuma (ed), *Comets and the Origin of Life*, ed, D. Reidel Publishing Co., Amsterdam

(2) Hoyle, F. and Wickramasinghe, N.C., 1981b. Space Travellers – The Bringers of Life, University College, Cardiff Press.

(3) Hoyle, F. and Wickramasinghe, N.C., 1983. Proofs that life is cosmic", *Mem.Inst.Fund.Studies, Sri Lanka*, No.1 (Revised edition published by World Scientific Publ. Co., 2018)

Journal of Cosmology, Vol. 30, No. 3, pp. 30030 - 30034

(4) Hoyle, F, and Wickramasinghe, N.C., 1985. Living Comets, University College, Cardiff Press

(5) Hoyle, F. and Wickramasinghe, N.C., 1986. The case for life as a cosmic phenomenon", *Nature,* **322,** 509

(6) Wallis, M.K. and Wickramasinghe, N.C., 2004. Interstellar transfer of microbiota, Mon. Not. R. Astron. Soc. 348, 52

(7) Wallis M.K., Wickramasinghe N.C., 2015. Rosetta Images of Comet 67P/Churyumov-Gerasimenko: Inferences from its Terrain and Structure. Astrobiol Outreach 3, 12

(8) Hoover, R.B. et al, 2022. ENAA and SEM Investigations of Carbonaceous Meteorites: Implications to the Distribution of Life and Biospheres, Academia Letters

(9) Wickramasinghe, N.C., Wickramasinghe, D.T. and Steele, E.J., 2019. Comets, Enceladus and panspermia, Astrophys Space Sci., 364, 205

(10) Wickramasinghe, N.C. 2022. Giant comet C/2014 UN271 (Bernardinelli-Bernstein) provides new evidence for cometary panspermia, *International Journal of Astronomy and Astrophysics,* 12, 1-6

(11) Hoyle, F. and Wickramasinghe, N.C., 1996. Life on Mars? (Clinical Press, Bristol)

(13) Clarke, Arthur C., 1999. Greetings, carbon-based bipeds!, Harper Collins, London

RELUCTANCE TO ADMIT WE ARE NOT ALONE AS AN INTELLIGENT LIFEFORM IN THE COSMOS

Chandra Wickramasinghe,[1,2,3,4] Gensuke Tokoro[2,4], Robert Temple[5] and Rudy Schild[6]

1. Buckingham Centre for Astrobiology, University of Buckingham, UK
2. Centre for Astrobiology, University of Ruhuna, Matara, Sri Lanka
3. National Institute of Fundamental Studies, Kandy, Sri Lanka
4. Institute for the Study of Panspermia and Astroeconomics, Gifu, Japan
5. History of Chinese Science and Culture Foundation, Conway Hall, London, UK
6. Center for Astrophysics, Harvard-Smithsonian, Cambridge, MA, USA

Email: ncwick@gmail.com

With an ever-increasing body of evidence from diverse scientific disciplines all pointing to the existence of alien life and alien intelligence on a cosmic scale, there has developed a growing tendency to maintain that we might still be alone as intelligent beings in the universe. This stubborn resistance to admit facts may well signal the end of our civilization.

Keywords: Exobiology, SETI, panspermia

1. Introduction

Yuri Milner, Russian billionaire who announced recently that he has relinquished his Russian citizenship, may have revived an ailing SETI (Search for Extraterrestrial Intelligence) programme with the injection of a 100 million dollars a few years ago. We are still waiting now for over a century for news of the first detection of an intelligent message from a cosmic neighbour. After such a long period even the genuine desire of humanity to discover an alien intelligence in all probability more advanced than us could be called to question.

It has recently been pointed out that such a message or messages may already have reached us, taking the form of microbial genetic codes that have already slipped into terrestrial biology (Slijepcevic, P. and Wickramasinghe, C., 2021). Temple and Wickramasinghe (2019) have also discussed a more specific model involving clouds of charged spinning dust grains that serve as conveyors of intelligent messages across the galaxy (see also: Temple, 2022). However, the more familiar expectation from SETI-type projects is the emergence of evidence for radio/microwave signals which has thus far been searched for over half a century with null, or at best dubious results.

2. Positive Evidence?

The SETI program (Search for Extra-Terrestrial Intelligence) began in 1960, supported first by NASA, and later by a few private or semi-private entrepreneurs. It is fair to say that any signs of success in this venture have been slow to come, and this was, of course, to have been expected. With the exception of the famous "Wow!" signal discovered in August 1977, there has been a deathly silence across all of the prospective sources and electromagnetic wavelengths that have been scanned.

In 1977, the "sound" of extra-terrestrials may have been heard for the first time – or so it was thought at the time. The famous Wow! Signal was detected by radio astronomer Jerry Ehman using Ohio State University's Big Ear Telescope. The telescope was a radio signal detector which, at the time, was pointed at a group of stars called Chi Sagittarii in the constellation Sagittarius. Whilst scanning the skies around the stars, Ehman recorded a 72 second burst of radio waves which he circled on his paper output with the word "Wow!" This record shown in Fig.1 is the source of the now famous "Wow!" signal – one that was never repeated. Over the following half a century the Wow! signal has tentatively been cited as possible evidence for the claim that we are not alone in the galaxy.

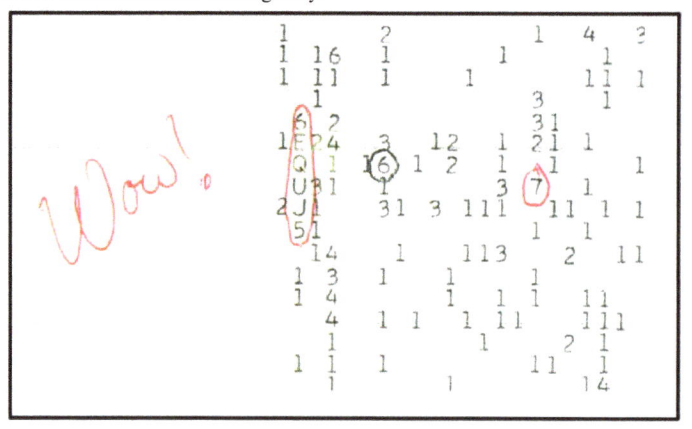

Fig.1. Ehman's Wow! signal

Recently, however, Antonio Paris (2017) attributed the Wow! signal to a pair of comets. The comets 266P/Christensen and 335P/Gibbs, which possessed extended comas of neutral hydrogen millions of kilometres in diameter, appear to coincide with the direction from which the signal came. This explanation may have some credibility because the Wow! signal, which was detected at 1420MHz, matches the radio frequency at which hydrogen naturally emits. Furthermore, Paris verified that these comets were in the "vicinity" at the time, and that the radio signals from 266/P Christensen matched those from the Wow! signal.

3. Case for Oumuamua being an alien craft

In 2017 another comet-like object Oumuamua some 200m long came into the solar system from interstellar space in a hyperbolic orbit and crossed the ecliptic plane on September 6th 2017. The object intercepted the ecliptic (solar system's orbital plane) from the direction of Vega, a massive bright star, some 25 light years away in the constellation of Lyra. After reaching perihelion (closest point to the Sun) the object evidently began leaving the solar system at a high relative speed, escaping the Sun's gravity.

Fig.2. Artist's impression of Comet Oumuamua that reached perihelion in a hyperbolic orbit in 2017. The aspect ratio of up to 10:1 is unlike that of any object seen in our own solar system. (Image Credit: European Southern Observatory / M. Kornmesser)

Since the surprise arrival of Comet Oumuamua in the solar system in 2017, some aspects of this object's behaviour have puzzled astronomers to the extent that a few have proposed it may be a product of alien intelligence (Loeb, 2022). One of us (NCW) and a few colleagues have, however, challenged these claims arguing the case for a natural cometary origin of the most secure data that relates to this object (Wickramasinghe, et al, 2018. 2019). However, Loeb and his colleagues have pointed out many aspects of the data relating to Oumuamua that may indeed be consistent with an "alien spaceship" hypothesis (Ginsburg et al 2018, Loeb, 2019, 2022).

4. Black Cloud hypothesis
Fred Hoyle's 1959 classic science fiction novel "Black Cloud" anticipated many of the future developments in science and astronomy regarding the persistence of organic molecules and even living entities throughout the galaxy (Hoyle, 1959). The novel tracks the progress of a giant black cloud that comes towards Earth and positions itself in front of the sun, thus causing widespread panic. The cloud is revealed to be an alien gaseous superorganism many times more intelligent than humans, one which is perhaps surprised to find intelligent life-forms on a solid planet like the Earth. The cloud suddenly removes itself from our solar system because it detects a signal from another black cloud in another part of the Galaxy and it rushes off to make a new friend. This allows sunlight to return to the Earth and thus humanity is saved. It is important to stress that Hoyle's novel was written before dusty complex plasmas were known, and so his black cloud has no plasma aspects in the way that we now know it must have.
Besides the power struggles that ensue between competing institutions and groups of scientists and astronomers who are challenged with the task of dealing with the situation, some aspects of the science involved chime with recent scientific discussions in science. In particular, recent arguments by two of us (Chandra

Wickramasinghe and Robert Temple) relating to the so-called Kordylewski Clouds are worthy of note (Temple and Wickramasinghe, 2019). They first point out that recent astronomical observations combined with dynamical simulations have led to a confirmation of the existence of stable dust clouds (Kordylewski Dust Clouds) at the Lagrange libration points of the Earth-Moon system. The diameter of a Kordylewski Cloud is estimated to be about 9 times the Earth's diameter, and the radius of the average grain is estimated at $\sim 3 \times 10^{-5}$cm, consistent with bacterial-type cells, with a mean separation of less than 1 cm. We have argued that such grains are most likely elongated and similar to rod-like bacteria, photoelectrically charged to a few eV, and would acquire a spin through collisions with neutral gas atoms, and thus could act as emitters and absorbers of longwave electromagnetic radiation. The entire Kordylewski Dust Cloud comprised of such particles would then have the potential to possess electromagnetic connectivity combined with an information storage/processing capacity akin to a form of intelligence – a gigantic superbrain, not unlike Fred Hoyle's fictional Black Cloud. In addition to these calculations, it should be noted that the clouds must be dusty complex plasma clouds, which have additional characteristics of their own which would further enhance the inevitability of self-organisation and emergence of intelligence, presumably a kind of 'super-AI' intelligence of such vast computing power that the human imagination is simply incapable of even imagining it.

5. Spontaneous Generation of Life or Panspermia?

A prerequisite for success of any form of SETI is the presence of technologically advanced lifeforms on a galactic scale. On the basis of the reigning dogma of spontaneous generation, life on Earth emerged as the result of an exceedingly improbable accident whereby the chemical building blocks of life (nucleotides/amino acids) assembled themselves in primitive oceans to form an "evolvable" microbial living system. Hoyle and one of us (Hoyle and Wickramasinghe, 1982) have argued that this would inevitably involve probability factors of the order of 1 in $10^{40,000}$ that cannot be easily grasped or understood within the context of a finite universe. (See also more recent arguments discussed by one of us (Wickramasinghe, 2013). If we argue that a "miracle" transcending these improbabilities *must* have happened on the Earth for the sole reason that we are here, then a similar miracle will not be repeated elsewhere. If, furthermore, we assume that the evolutionary progression from microbes to intelligent humans arose in a natural sequence of the most exceedingly improbable events, then it stands to reason that this entire process will not be repeated elsewhere, and we certainly would remain hopelessly alone in the cosmos.

This position of isolation ultimately stems from a philosophy that dates back to the 3rd century BCE. The Greek philosopher Aristotle proposed that life can emerge *spontaneously* from non-life on a planet like the Earth whenever the "right conditions" prevail. This basic concept, stretched beyond Aristotle's own intentions, forms the basis of the modern theory of spontaneous generation to which we have already referred and which involves improbabilities on a scale that cannot be bridged no matter how vast or how old a finite universe might be.

Let's now turn to evidence as it continues to unfold at the present time. How did life *really* arise in the first place? Not just on the Earth, or in our Milky Way galaxy, but anywhere in the Universe? Conventional science has steadfastly maintained that the Universe itself had a definite origin in time 13.8 billion years ago, a little in excess of

3 times the age of the Earth. With new data recently emerging from observations made with the James Webb telescope, serious potential flaws are emerging in this "standard" cosmology, possibly pointing to cosmologies which have an open timescale (Penrose, 2022). In such a cosmology one might well imagine a scheme that involves the "information" of life also to have an open timescale (Wickramasinghe *et al.*, 2023). The idea of life emerging spontaneously on planet Earth, whether by unknown physical processes or even by means of miraculous intervention, then becomes irrelevant.

6. Psychological Impediments?

Let us next ask the deeper question: do we, *Homo sapiens* in the year 2023, really want to come face to face with a superior extraterrestrial intelligence if such exists? Or would we rather turn away from even contemplating such a possibility? Coming to grips with the prospect of AI, which might possibly trump human intelligence in the very near future, is already sending shivers down our spines, so there may well be a deep psychological impediment to facing any prospect of encountering, or even admitting the fact of any extraterrestrial intelligence higher than our own prevailing in our midst, or entering our space of cosy comfort. At a more basic level, there may also be a subliminal resistance to entertaining the prospect that the Earth continues to be connected to a vast cosmic "ocean" of informationally rich extraterrestrial microorganisms and viruses, and this may well be one reason for holding back a paradigm shift that is long overdue (Wickramasinghe, 2023). Our autonomy as a species and our absolute independence and self-determination will then be called into serious question. The ongoing antagonism to the concept of cometary panspermia might thus have deeper roots than we are willing to accept.

7. An Ancestral Psychological Driving Force?

There may, however, be a redeeming feature in our behaviour over the past few decades. All the space missions undertaken in recent times spell out a single unavoidable cosmic truth: *Homo Sapiens,* in whatever way we emerged, appears to be hard-wired to seek out its cosmic origins, perhaps *intuitively* sensing that *we* cannot be alone and, more importantly, that we could not claim to be the most intelligent or the most technologically advanced life form in our cosmic neighbourhood. An inkling of the same realisation seems to show up even in the cave paintings of our stone-age ancestors perhaps as distant in time as 20,000 years ago, for instance in the famous Lascaux caves in the Dordogne region of southern France (See Fig.1).

Fig.3. Ancient cave painting showing constellation of Taurus and Pleiades in the Lascaux caves.

Martin Sweatman and Alistair Coombs studied the chemical makeup of the paint used in cave drawings and dated the art back 12,000 to 40,000 years. Next, they calculated the positions of stars were positioned at the times the art was created and concluded that many of the cave paintings mark the dates of significant comet sightings, and that the relative placements of stars correlated with stellar constellations that were visible at those times.

8. Scope of Panspermia

Given an access to the basic genetic units of life which we have argued are contained and carried in comets, and throughout interstellar space, the requirement for the emergence of creatures like ourselves (carbon-based bipeds) would be for the existence of rocky planets with water and an atmosphere generally similar to Earth (Wickramasinghe et al, 2019). This is of course to exclude higher levels of "Black Cloud"- type intelligence that may also be lurking around in our vicinity and seeking to communicate with us. Sticking to creatures modelled on ourselves, how many such planetary homes exist in the Galaxy, and beyond?

In 2009 NASA launched its orbiting Kepler telescope, which was specifically designed to discover planets which are the size of Earth. The detection process involved tracking down minute blinks (dimming) in the star's light when a planet transited periodically in front of it during its orbit. Extrapolating from the sample of present detections, the estimated total number of habitable planets in the galaxy is reckoned to be in excess of a billion and with a mean separation of only 10 parsecs (Kopparapu, 2013). Most of these habitable planets orbit very long-lived red-dwarf stars that are nearly twice as old as the sun. On the vast majority of these planets, it is possible that life may have begun, evolved to advanced levels of intelligence, and perhaps long since disappeared, their homes being engulfed in the expansion of their parent stars in the red giant phase.

Apart from the additional prospect of microbial SETI that we have discussed, radio, microwave and laser detections of potential ET signals are still the priorities being pursued. It is in such ventures that financial support from individuals like Yuri

Milner are still relevant and, of course, welcome. Such support could help to "buy" more telescope time, increase the range of wavelengths, enhance detector sensitivity and extend sky coverage. These new future developments have often been argued as necessary prerequisites if a SETI breakthrough within the foreseeable future is to be achieved. But it is worth stressing that a positive result from SETI would be contingent on the validity of the ideas of panspermia that we have developed for over four decades (Wickramasinghe et al., 2019). It is only the operation of panspermia, with the consequent widespread dispersal of information-carrying primitive life, that would lead to the emergence of intelligent lifeforms such as ourselves. How often does this happen?

The discovery of microorganisms occupying the harshest environments on Earth continues to provide indirect support for panspermia. Transfers of microbial life from one cosmic habitat to another requires endurance to space conditions for millions of years. The closest terrestrial analogue to this latter situation exists for microbes exposed to the natural radioactivity of the Earth. Quite remarkably, microbial survival under such conditions is now well documented. Dormant microorganisms in the guts of insects trapped in amber have been revived after 25-40 million years. All this goes to show that arguments used in the past to 'disprove' panspermia – an idea that had its beginnings in antiquity – independently in classical Greece, India and Egypt (Temple, 2007) - on the grounds of survivability during interstellar transport are seriously flawed. Another aspect of the circulation and transport of microbiota in space and the organisation of microbiota within cosmic clouds concerns the neglected aspect of *charge*. A paper is in preparation by two of us (Temple and Wickramasinghe) concerning the importance of charged microbiota, whether negative or positive. Just as there are two kinds of charged rain, positively charged at higher altitudes and negatively charged at lower altitudes, so there are two kinds of charged microbiota, positive and negative. Not only has this question not yet been discussed, but it appears that virologists have not even considered the effects if any which charge might have on the behaviour of viruses. (For instance, could some viruses be denatured or rendered less potent if subjected to charge reversals or repeated charge oscillations?)

Whilst Francis Crick and Leslie Orgel's idea of directed panspermia transfers the problem of the origin of life to another cosmic site, possibly invoking intelligent intervention (Crick and Orgel, 1973), modern advocates of panspermia have attempted to expand the domain in which cosmological abiogenesis *may* have occurred (Wickramasinghe et al., 2023). Once life has got started in the universe and evolved on an alien planet or planets the transference of the products of local evolution to other planets within reach becomes inevitable. Transference to other habitable planets or moons in the same system (eg between sites in our solar system) becomes more or less guaranteed over long enough periods of time. The same process can be repeated (via comet or asteroid collisions) to transfer genetic material carrying local evolutionary 'experience' to other distant molecular clouds containing nascent planetary systems. If every life-bearing planet transfers genes in this way to more than one other planetary system (say 1.1 on the average) with a characteristic time of 40My then the number of seeded planets after 9 billion years (lifetime of the galaxy) is $(1.1)^{9000/40} \sim 2 \times 10^9$ (Wallis and Wickramasinghe, 2004) Such a large number of 'infected' planets illustrates that Darwinian evolution, involving horizontal gene transfers, must operate not only on the Earth or within the confines of the solar system but on a truly galactic scale. Life throughout the galaxy on this picture would constitute a single connected biosphere.

9. Societal Resistance to Panspermia

The compelling spectroscopic evidence for panspermia, and consequently the widespread existence of biological dust in interstellar space and in comets, have been discussed for well over a decade and will not be repeated here (Hoyle and Wickramasinghe, 2000; Wickramasinghe, 2010). The existence of PAH's (polyaromatic hydrocarbons) both in interstellar clouds within our galaxy and in extragalactic sources have also been known for over 3 decades, and in our view, this has been incorrectly attributed to a non-biological origin. Biological aromatic molecules in the form of PAH's would be a natural result of the degradation of biological dust (bacteria and viruses) which we have argued makes up over 10 percent of carbon in interstellar space (Wickramasinghe, 2010).

Another phenomenon that is linked to PAH's and biological dust is the extended red emission (ERE) that has been observed in many extended astronomical sources (Witt & Schild 1988; Furton & Witt 1992; Perrin et al. 1995). These sources, including the Red Rectangle, emit radiation at red wavelengths that is readily explained on the basis of biological pigments. The biological aromatic model for ERE discussed by Hoyle & Wickramasinghe (1996) still remains the most reasonable explanation for this data. The competing at present in vogue, involving inorganically generated PAH's, are not easily justifiable in our view.

In addition to the growing body of astronomical evidence supporting panspermia over a broad front, there is also strong evidence for the continuing entry of bacterial material into the stratosphere of the Earth, amounting to some 20-200 million bacteria per square metre per day, a flux that will normally go unnoticed, but one that we cannot afford to ignore (Harris et al, 2001; Reche et al, 2018; Wickramasinghe et al, 2020).

In conclusion we stress that the prevailing societal resistance to panspermia is one that has to be overcome, for otherwise there will be a dangerous outcome for humanity. We have argued from 1979 onwards that pandemics of new viruses arriving from space pose a continuing threat to our planet. Many historic pandemics, including the 1918-1919 influenza pandemic that claimed 30 million lives, show all the signs of a space origin (Hoyle and Wickramasinghe, 1979). Similarly, other pandemics all the way down to the recent Covid-19 pandemic have signs of a cosmic content. In a recent paper we issued a "warning" as to how we might predict, and possibly prepare for, future eventualities of a similar kind (Qu and Wickramasinghe, 2020). Such warnings are ignored and when facts are rejected in favour of fashion and ideology there can be little doubt that the end is nigh.

What then of life outside Earth including SETI? Collisional ejection of life bearing rocks from a life-laden Earth to other nearby habitable sites in the solar system and beyond would appear to be more-or-less guaranteed. However, a reluctance to accepting available data including evidence for past and recent life on Mars and other habitable planets and moons in the solar system appears to be deep rooted in prejudice (Joseph *et al.*, 2023).

Another related matter concerns the objective and impartial assessment of the so-called UFO sighting reports. Here, perhaps understandably, prejudice runs at an even deeper level, and it is amply clear that such matters must be faced with honesty and courage. Humanity is perhaps not yet ready to admit that we are not alone – let alone invaded by alien life (Wickramasinghe and Tokoro, 2014; Grebennikova *et al.*, 2018; Harris *et al.*, 2002). Let us hope that it will not take another devastating pandemic, an H.G. Wells-type War of the Worlds, or even a Fred Hoyle-type Black Cloud to visit before reality and sanity dawns on *Homo Sapiens*.

REFERENCES

Furton, D.G. & Witt, A.N. 1992. Extended red emission from dust in planetary nebulae, *Astrophys. J.* 386, 587-603

Ginsburg, I., Lingam, M., and Loeb, A. 2018, Galactic Panspermia, *Astrophys.J.Lett.*, 2018, 868, L12

Harris, M.J., Wickramasinghe, N.C., Lloyd, D. et al, 2001. Detection of living cells in stratospheric samples, *Proc. SPIE*, 4495, 192-198

Hoyle, F. and Wickramasinghe, C. 1979. Diseases from Space, Chapman & Hall, London.

Hoyle, F. and Wickramasinghe, N.C., 1982. Evolution from Space, J.M.Dent & Sons, London.

Hoyle, F. & Wickramasinghe, N.C., 1996. Biofluorescence and the extended red emission in astrophysical sources, *Astrophys. Space Sci.* 235, 343

Joseph, R.G., Rizzo, V., Gibson, C.H. et al., 2023. Fossils on Mars? A "Cambrian Explosion" and "Burgess Shale" in Gale Crater? *Journal of Astrophysics & Aerospace Technology*, 11.01

Kopparapu, R.K. et al., 2013. Habitable zones around Main-Sequence stars: New estimates, *Astrophys.J.* 765, 131, 2013.

Loeb A. 2018. Six strange facts about the interstellar visitor 'Oumuamua'. *Scientific American*, November 20, 2018c.

Loeb, Avi, 2019. Extraterrestrial: The First Sign of Intelligent Life Beyond Earth. (Houghton Mifflin Harcourt, USA).

Paris, Antonio, 2017. Hydrogen line observations of cometary spectra at 1420MHZ, *Journal of the Washington Academy of Sciences, Vol 103, No.2*

Qu, J. and Wickramasinghe, N.C., 2020. The world should establish an early warning system for new viral infectious diseases by space-weather monitoring, MedComm, 3(1)

Reche, I., D'Orta, G., Mladenov, N. *et al.*, 2018. Deposition rates of viruses and bacteria above the atmospheric boundary layer, The ISME Journal, (https://doi.org/10.1038/s41396-017-0042-4)

Slijepcevic, P. and Wickramasinghe, C., 2021. Reconfiguring SETI in the microbial context: Panspermia as a solution to Fermi's paradox, *BioSystems*, 206, 104441

Steele E.J., Al-Mufti S., Augustyn K.K., *et al.*, 2018. Cause of Cambrian Explosion: Terrestrial or Cosmic? Prog. Biophys. Mol. Biol., 136: 3-23, https://doi.org/10.1016/j.pbiomolbio.2018.03.004

Steele E.J., Gorczynski R.M., Lindley R.A. et al., 2019. Lamarck and Panspermia - On the Efficient Spread of Living Systems Throughout the Cosmos. Prog. Biophys. Mol. Biol., 149: 10 -32. https://doi.org/10.1016/j.pbiomolbio.2019.08.010

Temple, R. and Wickramasinghe, C., 2019. "Kordylewski dust clouds: could they be cosmic "superbrains"? *Advances in Astrophysics, Vol. 4, No. 4, November 2019*. (Reprinted also as an appendix in Temple, 2022.)

Temple, R., 2007. The history of panspermia: astrophysical or metaphysical?, International Journal of Astrobiology, 62 (2), 169-180

Temple, R., 2022. A New Science of Heaven, Hodder & Stoughton, London, 2022.

Wallis, M.K. and Wickramasinghe, N.C., 2004. Interstellar transfer of microbiota, Mon.Not. R.A.S, 384(1), 52-61

Wallis, M.K. and Wickramasinghe, N.C., 2015. Rosetta Images of Comet 67P/Churyumov–Gerasimenko: Inferences from Its Terrain and Structure J.Astrobiol.Outreach, 3:1 DOI: 10.4172/2332-2519.1000127

Wickramasinghe, C. 2013. Simulation of Earth-based theory with lifeless results, BioScience, 63(2), 141-143

Wickramasinghe, C. (ed.), 2015. Vindication of Cosmic Biology, World Scientific Press, Singapore

Wickramasinghe, C., 2010. The astrobiological case for our cosmic ancestry, Int.J.Astrobiol., 9(2), 119.

Wickramasinghe, C., 2022. Panspermia verus Abiogenesis: A clash of cultures, Journal of Scientific Exploration, 36(1), 121

Wickramasinghe, J.T., Wickramasinghe, N.C. and Napier, W.M., 2010. Comets and the Origin of Life (World Scientific Pub. Singapore)

Wickramasinghe, N.C. and Tokoro, G., 2014. Life as a Cosmic Phenomenon: The Socio-Economic Control of a Scientific Paradigm, J. Astrobiol.Outreach, 2(2) 1000113

Journal of Cosmology, Vol. 30, No. 4, pp. 30040 - 30053

Wickramasinghe, N.C., 2012. DNA sequencing and predictions of the cosmic theory of life, Astrophysics and Space Science, 7 September 2012

Qu, J. and Wickramasinghe, N.C., 2020. The world should establish an early warning systemfor new viral infectious diseases by space-weather monitoring, MedComm. 2020;1–4.

Wickramasinghe, N.C., Steele, E.J. Wallis, D.H. *et al.* 2018. Oumuamua (A/2017U1) – a confirmation of links between galactic planetary systems. Advances in Astrophysics, Vol. 3, No. 1, February 2018

Wickramasinghe, N.C., Steele, E.J. Wallis, D.H. *et al.*, 2018. Oumuamua (A/2017U1) – a confirmation of links between galactic planetary systems. Advances in Astrophysics, Vol. 3, No. 1, February 2018

Wickramasinghe, N.C., Tokoro, G. and Temple, R., 2021. Intelligent messages in bacterial DNA- a sequel to SETI? Advances in Astrophysics, Vol. 6, No. 1, February 2021

Wickramasinghe, N.C., Wickramasinghe, D.T. and Steele, E.J., 2019. Ouamuamua (A/2017U1), Panspermia, and Intelligent Life in the Universe, Advances in Astrophysics, Vol. 4, No. 3, August 2019

Wickramasinghe, N.C., Wickramasinghe, J.T. and Wallis, D.H., 2018. The size and albedo of the object of interstellar origin A/2017U1, Advances in Astrophysics, Vol. 3, No. 2, May 2018

Wickramasinghe, N.C., Narlikar, J.V. and Tokoro, G., 2023. Cosmology and the Origins of Life, Journal of Cosmology, Volume 30.
https://thejournalofcosmology.com/Wickramasinghe.pdf

Witt, A. & Schild, R., 1988. Hydrogenated amorphous carbon grains in reflection nebulae, Astrophys. J. 325, 837-845.

Wickramasinghe, N.C., Maganarachchi, D., Temple, R. et al, 2020. The Search for Bacteria and Viruses in the Stratosphere The search for bacteria and viruses in the stratosphere, Advances in Astrophysics, 5(2),
https://dx.doi.org/10.22606/adap.2020.52003

Journal of Cosmology, Vol. 30, No. 5, pp. 30060 - 30071

The Second Copernican Revolution

Chandra Wickramasinghe,[1,2,3,4], Rudy Schild[5] and J. H. (Cass) Forrington[6]

1. Buckingham Centre for Astrobiology, University of Buckingham, UK
2. Centre for Astrobiology, University of Ruhuna, Matara, Sri Lanka
3. National Institute of Fundamental Studies, Kandy, Sri Lanka
4. Institute for the Study of Panspermia and Astroeconomics, Gifu, Japan
5. Center for Astrophysics, Harvard-Smithsonian, Cambridge, MA, USA
6. United States Merchant Marine Academy, Kings Point, N. Y., Cum Laude, 1972

The recent discovery by the James Webb Space Telescope of organic molecules possibly related to life in a galaxy at redshift z=12.4 may well signal a concluding phase of the second Copernican revolution, thus removing the Earth from the centre and focus of biology and charting a new course in our understanding of the universe, and concluding a process that began 4 decades ago.

Keywords: Panspermia, cosmology, astrobiology, origin of life

1. Introduction

*"It appears to me to be a fully correct scientific procedure,
if all our attempts fail to cause the production of organisms
from non-living matter, to raise the question whether life
has ever arisen, whether it is not just as old as matter itself..."*-

Herman von Helmholtz (1821-1894)

Forty years ago, Fred Hoyle and one of the present authors (CW) challenged what was essentially regarded as the bedrock of Western culture and science – the origin and evolution of life interpreted in a narrow terrestrial setting (1-10). The idea of life emerging on Earth spontaneously from inorganic material constituted the central core of Aristotelean philosophy and had dominated western thinking for over two millennia. Yet a veritable tide of new facts from astronomy and molecular biology that emerged after the dawn of the space age in the 1960's led Hoyle, Wickramasinghe and their collaborators to seriously question this reigning dogma.

From the early 1980's Hoyle, Wickramasinghe and others followed a path that had seemed too daunting for their predecessors in former times, forging what could be seen as a merger of astronomy and biology, and thus leading to the birth of the new discipline of astrobiology (5-10). This chosen path was beset with societal disapproval on a scale that had not been witnessed for a long time. However, encouragement to continue and not to be deflected from an intended path was derived not from individuals as such, but from new scientific facts that still continues to emerge from diverse disciplines – geology, biology and astronomy itself.

A related paradigm that took centuries to overcome also had Aristotelean roots – the premise that the Earth was the centre of the solar system and centre of the Universe itself. The Copernican revolution in astronomy started with Galileo and Copernicus and was concluded with Kepler and Newton, spanning the time interval 1500-1700CE, nearly 200 years.

38 *Convergence to Cosmicrobia: The Final Acceptance of Life as a Cosmic Phenomenon*

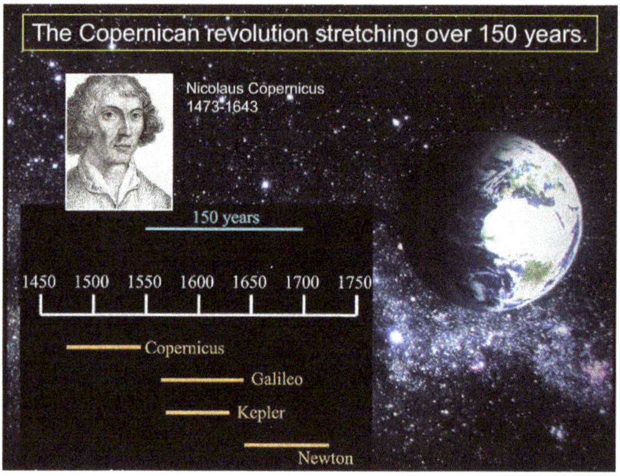

Fig. 1. Trajectory of the first Copernican revolution that removed Earth from the centre of the solar system and Universe.

This "first" so-called Copernican revolution displaced the Earth from its earlier hallowed status of centrality in the cosmos. The developments in astrobiology and cosmology over the past four decades could be seen to define the second and final part of the same Copernican revolution, one that now seeks to remove Earth from the centre of biology and advance the proposition that *life is a cosmic phenomenon*.

2. Modern resurrection of Panspermia

Panspermia is an ancient idea that has had a chequered history from its first emergence in the canon of pre-Socratic philosophy with the pronouncements of Anaxoragas in the 5th century BCE. Nearly two centuries later, the far more influential Greek Philosopher Aristotle effectively took control of philosophy over an exceedingly broad front, and consequently his rejection of panspermia and support for the rival theory of spontaneous generation dominated Western thinking in this area almost to the present time.

Panspermia of course had many staunch advocates who emerged sporadically through the centuries (Wickramasinghe et al, (9); Wainwright and Wickramasinghe, (10)). Perhaps the most notable amongst them was the French biologist Louis Pasteur who challenged Aristotelean spontaneous generation over a century ago (11). From Pasteur's laboratory studies on the growth of microorganisms in broths he confidently concluded with the strongly positive declaration *"Omne vivum ex vivo"*, which means 'Life [is] from Life' in 1864. This claim was of course vigorously refuted by most biologists in the early 20th century, but as the decades progressed and with the emergence of new evidence from biology and astronomy, one of the present authors (CW) together with Hoyle and his collaborators were able to reaffirm the same position.

As with the first Copernican revolution, and indeed all earlier major paradigms throughout history, "wrong" ideas about the nature of the world that have become entrenched and been in currency for extended periods have always been difficult to displace. The old ideas are passionately adhered to and vigorously defended until, with the advent of new facts, they must of necessity eventually come to be overturned. The same, we believe, will undoubtedly be true today in relation to all the "big ideas" about the world. One such big idea is Aristotle's theory of the spontaneous generation of life. This theory led naturally to the firmly entrenched modern belief that life is a purely terrestrial affair – a trivial and obvious assembly of atoms – atoms that were of course synthesized through nuclear processes in the deep interiors of stars. The first cells to "emerge" spontaneously from such atoms, cells from which all the rest of life eventually evolved, is assumed to arise as a result of chemicals coming together *spontaneously* to form "life" in some location on Earth – margins of oceans, lakes, ponds or geothermal vents. Yet, it goes without saying there is, and never was, any substantive evidence to support such a belief.

What is usually ignored, or at any rate consistently glossed over, is the astronomically vast information content (in the form of the *specific arrangements* of the relevant monomers) that is involved in formation of even the simplest living cell (2). In present-day biology, the so called "Shannon information" contained in enzymes—the specific arrangements of amino acids in folded chains—is crucial for the functioning of life, and this information is transmitted by means of the coded ordering of nucleotides in DNA. In a hypothetical RNA world, that may have predated the DNA-protein world, RNA is normally posited to serve a dual role as both enzyme and genetic transmitter. If a few ribozymes are regarded as precursors to all life, one could attempt to make an estimate of the probability of the assembly of a simple ribozyme composed of 300 bases. This probability turns out to be 1 in 4^{300}, which is equivalent to 1 in 10^{80}, which can hardly be supposed to happen even once in the entire 13.7-billion-year history of the universe.

The biggest problem relating to the origin of life is to overcome an astronomical or super-astronomical improbability hurdle that demands stochastic resources not available on the Earth. The Earth of course is an open system inextricably linked to the external universe by continually receiving material inputs in the form of cometary dust, meteorites etc; and as we now know, following studies with the Kepler space telescope, the solar system is within easy reach/connection of billions of other habitable planets in the galaxy. There is thus no logical requirement any more to insist that life necessarily started on Earth.

The conventional and societally sanctioned idea of life originating *de novo* on Earth can only be defended by placing the Earth in a unique and special position in relation to the occurrence of perhaps the most improbable of cosmic events. In the exceedingly unreal circumstance that no "bigger" probability space is available for such an event we might even be forced to adopt an anthropic argument in this context and say: "We are here on the Earth, so life *must* have started here no matter how improbable that might be".

Since the discovery of DNA by Francis Crick in the 1950's, combined with the discovery and sequencing of enzymes, the informational hurdle to overcome in transiting from life from non-life came into sharp focus (2). The precise ordering of nucleotides in DNA, or correspondingly the arrangement of amino acids in enzymes, poses a difficulty that can only be resolved if we are permitted to extend the domain (or domains) for life's origin to encompass cosmic or even cosmological dimensions. Within such a domain of cosmic (or even infinite) proportions, the superastronomical informational hurdle for starting life *could* be somehow overcome.

Removing the Earth from the centre of biology in such a way would arguably constitute the second and ultimate Copernican revolution.

When organic molecules were first discovered in interstellar clouds in the 1960's it might be said that the die was cast for abandoning the Earth-centred view of life. Contemporaneously with the discovery of interstellar organic molecules of ever-increasing complexity, a major paradigm shift in astronomy was being spear-headed by Fred Hoyle and one of the present authors (4-6). At this time the strongly held astronomical opinion was that interstellar dust was comprised of ice grains that condensed in interstellar clouds. A combination of mathematical modeling and astronomical observations led over the decade 1962-1972 to a transition from the old ice particle model of interstellar dust to a model involving mixtures of carbonaceous and siliceous grains. From 1974 onwards, the interstellar dust models began to include a large component in the form of organic polymers – eg polyoxymethyline, polysaccharides (5, 6).

In the late 1970's the time seemed right to confront the long-held view that life originated on the Earth by expanding the canvas for life's origin to span the biggest available galactic distance scales. Laboratory evidence demonstrating the incredible space-survival attributes of bacteria and viruses was accumulating to support such a model. It could be argued then that once life had originated in a cosmic setting its subsequent spread and perpetuation was inevitable. These ideas were tantamount to a revival and radical re-casting of the ancient idea of panspermia, with comets playing a crucial role in the amplification and dispersal of life on a galactic scale (6-10). The new science of astrobiology was born at about the same time (7).

If habitable planets were commonplace, which we did not know at the time, but now know, the entire galaxy will become a single connected biosphere on a relatively short timescale (9). The time scale could be as short as 240 Myr which is the period of rotation of the solar system around the centre of the Galaxy, during which mixing of biological material between planetary systems would inevitably occur. The adoption of such a point of view regarding the astronomical origins of life would surely open up new vistas of research in astronomy as well as biology.

This is the "Brave New World" approach that still continues to be resisted by many in the scientific community for reasons that lie well outside the realm of science. Supportive scientific evidence for the idea itself has accumulated steadily over the past 4 decades from widely different areas of science – astronomy, biology, geology. However, each individual piece of evidence supporting such a cosmic origin of life can be interpreted, if one so chooses, in a conservative manner that might, albeit imperfectly, preserve the *status quo*. We would therefore quite naturally expect to see a series of major confrontations with authority in an attempt to deny or delay an inevitable conclusion. This is precisely what happened.

3. Data from interstellar dust

An astronomical dataset that has not been explained satisfactorily for over a century relates to the so-called diffuse interstellar absorption bands (DIB's) in the optical spectra of reddened stars. The strongest of these is centred at the wavelength 4430A and has a half-width of ~ 30A (12, 13). A possible solution to this 100-year old mystery is most likely to be connected with cosmic biology, a proposition that had been considered inadmissible from the very outset. This involves the behaviour of fragmentation products of biology existing in interstellar space under various conditions of excitation in their electronic configurations. A possible candidate in this category was proposed by F.M. Johnson in the form of molecule magnesium tetrabenzo-porphyrine (12, 13), a molecule related to chlorophyll that is of course a molecule crucial for photosynthetic biology. It is no surprise in this context that the overwhelming biomass of the Earth is in the form of chlorophyll-containing plants, algae and photosynthetic biology making up over 99.999% of the total.

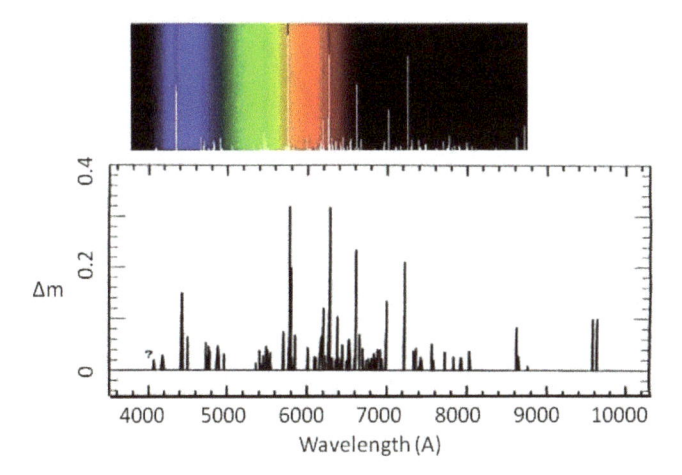

Fig.2. The diffuse interstellar bands with half widths 2-30A are distributed over the entire visual waveband. They are correlated with interstellar extinction but have defied identification for nearly 100 yr (see ref.12, 13).

The next confrontation with authority involved the interpretation of the interstellar extinction curve – the precise manner in which the continuum radiation of distant stars is scattered and absorbed by interstellar dust over a wide wavelength range from the infrared to the far ultraviolet. From the early 1970's onward, one of the present authors (NCW) had been working on the problem of interstellar dust for over a decade and was becoming increasingly uncomfortable with the attempts that were made thus far to theoretically model the data with inorganic dust models – particularly ice particles (4, 5, 6, 12). The fits of all such models to astronomical data were always less than perfect, and the assumptions required to obtain even imperfect fits were often too arbitrary and *ad hoc*.

In the late 1970's the emphasis shifted away from "dirty ice grains" to the possibility that a significant fraction of interstellar dust had an organic composition with biological connotation. With such models one of us (CW) was able to obtain impressively close fits to astronomical data and with a minimum of free parameters to fit.

One prediction of the organic model of interstellar dust was that the mid-infrared spectrum of an infrared source seen through a few kiloparsecs of dust obscuration should reveal a spectrum of bacterial material. In 1981 this prediction was dramatically verified at the Anglo-Australian Telescope by D.A. Allen and Dayal Wickramasinghe (14) (CW's brother) to a degree of precision that seemed stunning. This crucial fit is shown in the left-hand panel of Fig.3.

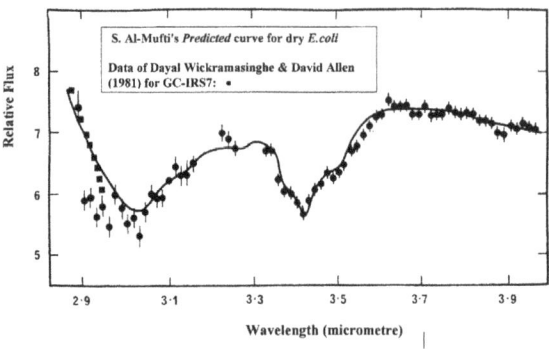

Fig 3. The first detailed observations of the Galactic centre infrared source GC-IRS7 (Allen and Wickramasinghe, 1981 (14) compared with earlier laboratory spectral data for dehydrated bacteria.

Later predictions for cosmic dust models with a biological provenance included an ultraviolet peak in the interstellar extinction curve at the wavelength 2175A which Hoyle and one of us (CW) attributed to bicyclic aromatic molecules (eg pyran rings in biology) (15). This feature was thus elegantly explained on the basis of biological dust models. Subsequently the prevalence of a similar UV dust feature in galactic dust and also dust in distant galaxies lent strong support to the concept of life as a cosmic phenomenon. Fig. 4 shows the appearance of this UV extinction feature in our galaxy as well as in external galaxies up to relatively high redshift values, implying in our view, that biological dust was present on a vast extragalactic scale (16, 17).

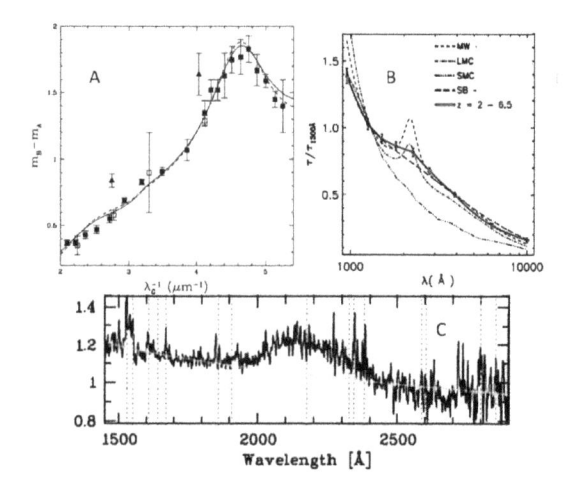

Fig. 4. Ultraviolet extinction curves for high redshift galaxies showing the 2200A feature which we attribute to biology. A: Motta et al (16) galaxy at z=0.83 (14); B: Stack of "normalised" extinction curves compiled by Scoville et al (17) for our galaxy (MW), LMC, for starburst galaxies (SB) and for the average of all high redshift galaxies with z in the range 2-6.5.

4. PAH molecules in space and Extended Red Emission of interstellar dust

The existence of PAH's (polyaromatic hydrocarbons) both in interstellar clouds within our galaxy and in extragalactic sources have also been known for over 3 decades, and in our view, this has been incorrectly attributed to a non-biological origin. Biological aromatic molecules in the form of PAH's would be a natural result of the degradation of biological dust (bacteria and viruses) which we have argued makes up over 10 percent of carbon in interstellar space (7). Also linked to PAH's is the phenomenon of extended red emission (ERE) that has been observed in many extended astronomical sources (Witt & Schild 1988 (18); Furton & Witt 1992 (19); Perrin et al. 1995 (20)). These sources, including the Red Rectangle, emit radiation at red wavelengths that is readily explained on the basis of biological pigments. The biological aromatic model still remains the most reasonable explanation for this entire data set (3, 19, 20).

Competing models based on emission by compact PAH systems, derived non-biologically, are not as satisfactory, as is evident in Fig. 5. Hexa-peri-benzocoronene is one of a class of compact polyaromatic hydrocarbons that have been discussed in this connection. However, the width and central wavelength of its fluorescent emission show a significant mismatch.

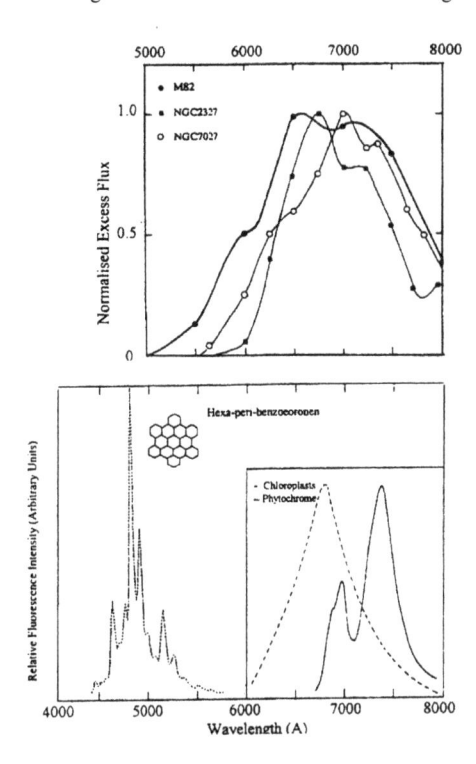

Fig. 5 Extended red emission (ERE). The points in the top panel show normalised excess flux over scattering continua from data of Furton and Witt (19) and Perrin et al. (20). The bottom right panel shows relative fluorescence intensity of spinach chloroplasts at a temperature of 77 K. The dashed curve is the relative

fluorescence spectrum of phytochrome. The bottom left panel is the fluorescence spectrum of hexa-peri-benzocoronene.

PAH (polyaromatic hydrocarbon) absorption bands are found in dust clouds in the Galaxy and in external galaxies up to high redshift values. The persistent PAH features in 13 starburst galaxies studied by Brandl et al (21) are shown in Fig. 6. Although the general preference among astronomers has been to interpret these features as the products of non-biological chemical processes such an explanation is far-fetched compared to the idea that they are the degradation products of biology.

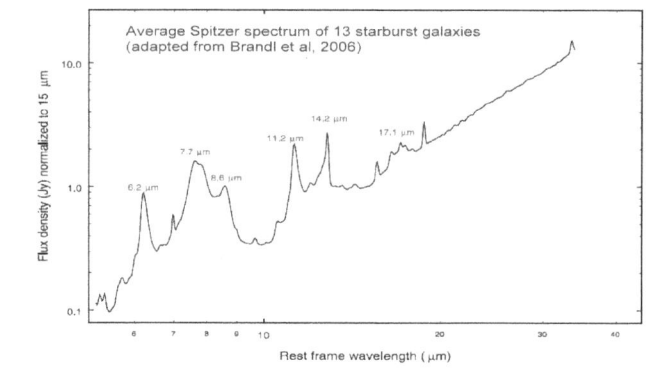

Fig. 6. Averaged *Spitzer* IRS spectrum of 13 starburst galaxies (21).

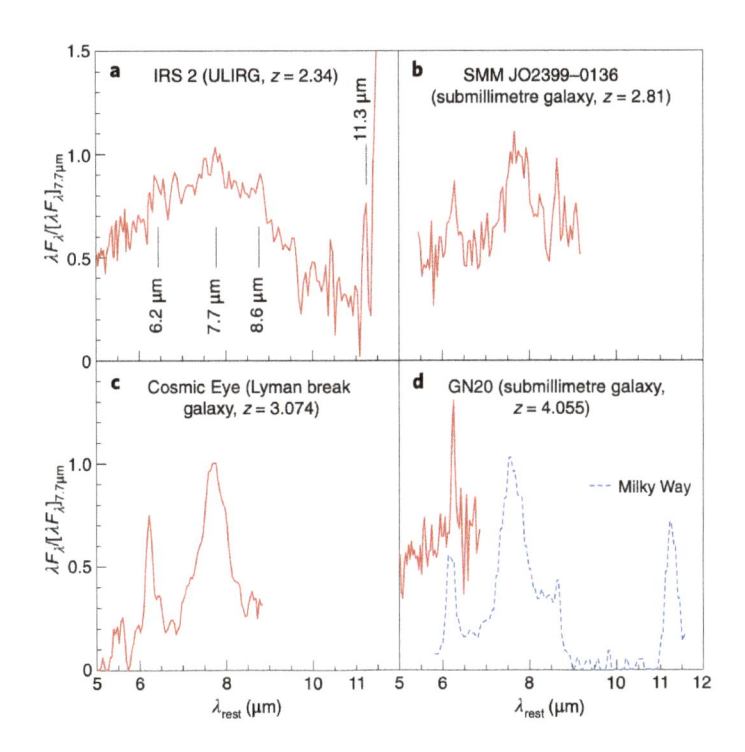

Fig. 7. Spitzer/IrS rest-frame spectra of galaxies at various redshifts (a) ULirG irS 2 ($z = 2.34$); (b) Submillimetre galaxy SMM J02399−0136 ($z = 2.81$); (c), Lyman break galaxy Cosmic Eye ($z = 3.074$); (d), Submillimetre galaxy GN20 ($z = 4.055$). Also shown in (d) is the I SOPHOT spectrum of the Galactic diffuse ISM (blue dashed line).

Fig.7 shows the PAH spectra of several high redshift galaxies recently compiled by Li (22), with wavelengths reduced to rest frame values. Also shown in the blue curve (7d) is the diffuse PAH emission from Milky Way. The vast quantities of PAH-type molecules and other organic molecules on the Earth are, in our view, unequivocally the degradation products of biology. To resist the same conclusion – even provisionally - for similar carbonaceous molecules that have been discovered in astronomy (galactic as well as extragalactic) for over 4 decades is unnecessary, to say the least, and is the reflection of the strongly entrenched pre-Copernican and strictly Aristotelean view of cosmic biology.

The most recent claim of "evidence of pre-biology" involves the detection of the ion CH3+ in a protoplanetary disk, and is yet another case in point (Spiker et at, 23). (See also: https://www.esa.int/Science_Exploration/Space_Science/Webb/Webb_makes_first_detection_of_crucial_carbon_molecule_in_a_planet-forming_disc). This new discovery is being hailed as signalling the operation of the super-astronomically improbable process of spontaneous generation, but there is no evidence whatsoever to support the causal link:

Organic molecules in protoplanetary disk → origin of life.

This most recent JWST discovery, along with the other features discussed earlier, must all be regarded as representing the detritus of life, in various stages of its inevitable degradation under cosmic conditions. It is, however, inevitable that such detritus will be available for the generation of biospheres on newly forming habitable planets, just as detritus of terrestrial life nurtures new life. This would be more relevant in the case of complex molecules that are involved in the ERE phenomenon (Witt and Schild (18)) and also the well-known diffuse interstellar features (12, 13) that have defied understanding for close on a century. However, we must stress that new life requires a full superastronomical content of Shannon information that must be supplied by an eternally present component of intact RNA and/or DNA which must in all likelihood be regarded as ever present in the cosmos. Such information must be, in the words of Helmholtz, "just as old as matter itself".

5. Completion of the Second Copernican Revolution

The overwhelming body of evidence for the enormous abundance of organic molecules in space – in comets, interstellar space and in external galaxies cannot fail to impress an impartial observer that we are facing an age-old cultural conflict – life as a cosmic phenomenon *vs* life as being purely terrestrial. Needless to say, all this evidence and in particular and the verification of our predictions of a biological model of interstellar dust (eg. Fig.3) continues to be vigorously resisted within mainstream astronomy. A non-biological explanation of the points in Fig. 3 would require inorganically formed organic molecules to possess functional groups that fortuitously mimic biology against the most impossible odds. Similar arguments have also been advanced for every subsequent discovery of astronomical spectra that matched biological material for example the data in Figs. 4-7.

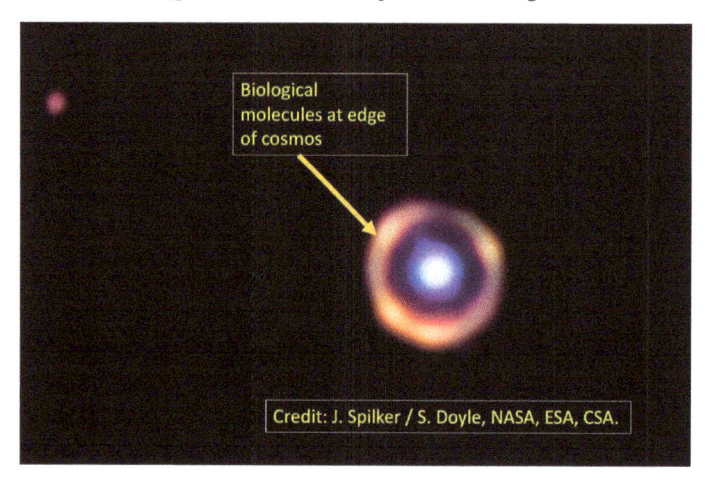

Fig.8. Einstein ring image of galaxy SPT0418-47, within which poyatomic hydrocarbon molecules were detected (*Image: J. Spilker / S. Doyle, NASA, ESA, CSA*)

Perhaps the most dramatic result to come from the James Webb Space Telescope (JWST) is the detection of complex organic molecules in a galaxy 12.3 billion light-years away (SPT0418-47) that was fortuitously gravitationally lensed by a foreground galaxy (23). The new observation indicated that SPT0418-47 was rich in heavy elements as well as complex organic molecules (PAH's). Such life-related molecules now reside a galaxy that formed when the universe was less than 10% of its current age. The light bearing spectroscopic signatures of these molecules began its journey scarcely 1.5 billion years after the putative Big Bang – if that indeed was the ultimate origin of the Universe. As we have seen in this review, evidence of complex organic molecules in external galaxies has accumulated over the past 3 decades but not beyond redshifts greater than z=4. The astronomical orthodoxy has maintained without proof that such molecules must necessarily be the result of non-biological chemical processes involving gas-phase reactions in interstellar clouds, but well-attested evidence now clearly supports the interpretation that Hoyle and one of us had suggested 40 years ago - namely cosmic organic molecules can plausibly arise only from the degradation of life itself – bacteria and viruses degrading under conditions prevailing in space.

With so many different and independent data sets and lines of evidence, all pointing to life as a cosmic phenomenon as a preferred option, and particularly with the new detection of organics in a galaxy uncomfortably close to the putative Big Bang event, the case for an Earth-centred origin of life is beginning to look ever more insecure. Life is a cosmic phenomenon, with the stark possibility that must be left open – both the Universe and life within it had no beginning!

The ultimate Copernican revolution has been completed and it will be the privilege of future generations to reap the benefits of this momentous transition.

Since the completion of this volume an exoplanet K2-18b replete with hot water oceans was discovered at a distance of 124 light years from the solar system and was found to have evidence of the biomolecule dimethyl sulfide, a molecule which is only known to be produced on the Earth by phytoplankton (Madhusudhan, N. *et al* 2023 *Ap.J.Lett,* **956** L13, **DOI** 10.3847/2041-8213/acf577)

REFERENCES

(1) Hoyle, F. and Wickramasinghe, N.C., 1983. Proofs that life is cosmic, *Mem.Inst.Fund.Studies, Sri Lanka,* No.1.

(2) Hoyle, F. and Wickramasinghe, N.C., 1982. Evolution from Space, J.M. Dent, London.

(3) Hoyle, F. and Wickramasinghe, N.C., 1983. Bacterial life in space, *Nature,* 306, 1983.

(4) Hoyle, F. and Wickramasinghe, N,C., 1982. From grains to bacteria, University College, Cardiff Press (Cardiff, UK)

(5) Hoyle, F. and Wickramasinghe, N.C., 1986. The case for life as a cosmic phenomenon", *Nature,* **322,** 509

(6) Hoyle, F. and Wickramasinghe, N.C., 1991. *The Theory of Cosmic Grains.* Dordrecht: Kluwer Academic Press.

(7) Wickramasinghe, C., 2010. The astrobiological case for our cosmic ancestry, International Journal of Astrobiology 9 (2) : 119–129

(8) Wickramasinghe, C., 2015. The search for our cosmic ancestry, World Scientific, Singapore

(9) Wickramasinghe, C., Wickramasinghe, K. and Tokoro, G., 2019. Our cosmic ancestry in the stars (Bear & Co., USA)

(10) Wainwright, M. and Wickramasinghe, C., 2022. Life comes from space, World Scientific, Singapore

(11) Pasteur, Louis, 1857. Comptes rendus de l'Académie des Sciences, 45, 913.

(12) Wickramasinghe, N.C., 1967. Interstellar Grains, London: Chapman & Hall

(13) Johnson, F.M., 1965. Diffuse interstellar lines and the chemical characterisation of interstellar dust, in Interstellar Grains (ed. J.M. Greenberg & T.P. Roark) NASA SP-140

(14) Allen, D.A. and Wickramasinghe, D.T., 1981. Diffuse interstellar absorption bands between 2.9 and 4.0 μm. *Nature*, **294**, pp. 239-240.

(15) Hoyle, F. and Wickramasinghe, N.C., 1977. Identification of the λ2200A interstellar absorption feature. *Nature*, **270**, pp. 323-324.

(16) Motta, V. et al., 2002. Detection of the 2175A Extinction Feature at $z = 0.83$. *Astrophys. J.*, **574**, pp. 719-725.

(17) Scoville et al (2015). Dust attenuation in high redshift galaxies: "diamonds in the sky", Astrophys.J., 800, 108

(18) Witt, A. & Schild, R., 1988. Hydrogenated amorphous carbon grains in reflection nebulae, *Astrophys. J.* 325, 837-845

(19) Furton, D.G. & Witt, A.N. 1992. Extended red emission from dust in planetary nebulae, *Astrophys. J.* 386, 587-603

(20) Perrin, J.M. et al. 1995. Observation of extended red emission (ERE) in the halo of M82 arXiv preprint astro-ph/9512046, 1995 - arxiv.org

(21) Brandt et al 2006. The mid-infrared properties of starburst galaxies from Spitzer-IRS spectroscopy, Astrophys. J., 653, 1129

(22) Li, Aigen, 2020. Spitzer's perspective of polycyclic aromatic hydrocarbons in galaxies Nature Astronomy Vol.4, 339-351

(23) Spilker, J.S., Phadke, K.A, Aravena, M. et al, 2023. Spatial variations in aromatic hydrocarbon emission in a dust-rich galaxy, *Nature*, **618**, 708–711

SEARCH FOR ALIENS, AND UFO'S

Chandra Wickramasinghe,[1,2,3,4] Rudy Schild[5], Gensuke Tokoro[2,3], Robert Temple[4] and J. H. (Cass) Forrington[6]

1. Buckingham Centre for Astrobiology, University of Buckingham, UK
2. Centre for Astrobiology, University of Ruhuna, Matara, Sri Lanka
3. National Institute of Fundamental Studies, Kandy, Sri Lanka
4. History of Chinese Science and Culture Foundation, London, UK
5. Center for Astrophysics, Harvard-Smithsonian, Cambridge, MA, USA
6. United States Merchant Marine Academy, Kings Point, N. Y., Cum Laude, 1972

The widespread existence of primitive life in the form of bacteria and viruses in the universe combined with the large numbers of habitable planets that are being discovered, leads to the serious possibility that intelligent life could be widespread throughout the cosmos. Discovering such alien intelligence in our vicinity continues to pose a challenge.

Keywords: Exobiology, Aliens, UFO's

"As far as these suns and moons revolve, shedding their light in space, so far extends the thousand-fold world system. In it there are a thousand suns, a thousand moons, a thousand inhabited Earths and a thousand heavenly bodies .... "

Anguttara Nikaya Sutta - Siddhartha Gautama Buddha
(circa 5th century BCE)

1. Introduction

We humans, *Homo sapiens sapiens*, since our first emergence some 300,000 years ago, have always sought to discover alien life in natural phenomena that we failed to understand. Today, in 2023, we are in essence no different. A lingering belief in the supernatural falls in this category, with an omniscient God, a pantheon of gods and goddesses or an evanescent world of spirits to explain the forces of nature continues unabated from the time of our stone-age ancestors to modern humans in the age of science. When these forces of nature eventually came to be understood, and in some instances even tamed, the quest for aliens did not cease; it simply took other forms.

A readiness to believe in Aliens and UFO's pervades popular culture of the 21st century, despite the tenuous nature of much of the evidence that is adduced in its support. The desire to believe in alien life is also encapsulated in the ever-increasing popularity of science fiction fantasising about aliens and alien life on other worlds. The famous radio broadcast as far back as 1938 of a dramatization of H.G. Well's *"War of the Worlds"* caused panic in the streets of New York, exemplifying again mankind's readiness to receive news that there is life out there – even intelligent life – on our neighbouring planet Mars. Notwithstanding such an instinct, official governmental bodies and the institutions of science in the 21st century behave differently. An attitude of self-righteous conservatism still prevails. Acceptance of even the strongest evidence of primitive life on Mars or elsewhere in the Universe seems hard to achieve in a climate where extraterrestrial life is regarded as an "extraordinary claim for which extraordinary evidence" is demanded. This phenomenon is clearly documented in the earlier papers in this Journal (Volume 30, and earlier).

2. Primitive alien life and intelligence

The ingress of alien microbial life onto our planet, whether dead or alive should not by any rational argument be perceived as a cause for concern. From stratospheric probes launched by the Indian Space Research Organisation (ISRO) as far back as 2001 we discovered evidence for some 20,000-200,000 bacteria per square metre per day, falling from the stratosphere at 41km (Harris et al, 2001 (1)). Such a flux when it becomes mixed with terrestrial microbiota will normally go unnoticed, but this is clearly a component of cosmic alien microbial life that we cannot afford to ignore. We have argued earlier that such extraterrestrial microbes might even be carriers of coded genetic messages that may need to be deciphered if intelligent messages are to be discovered (Slijepcevic and Wickramasinghe (2); Temple and Wickramasinghe (3);Wickramasinghe et al (4)). We have also connected this in-fall of bacteria from cosmic sources as a source of evolutionary potential (5), we have here a process that has continued throughout geological time from the moment of the first bacteria arriving on the Earth some 4.2 billion years ago.

Unlike the prospect of discovering alien intelligence that might be justifiably viewed with apprehension by Governments of the world, the humblest of microbial life-forms occurring on a cosmic scale would not, or at any rate should not, constitute any serious threat. Neither would the discovery of alien microbes impinge on any issues of national sovereignty or defence, nor challenge our long-cherished position as the dominant life-form in our corner of the Universe. We have already discussed these matters in our earlier papers in the present volume (Vol. 30) of JoC. Aliens on other planets and further afield in the cosmos may, however, be viewed with apprehension. In particular the possible evolution of these alien life forms into creatures with intelligence that could even surpass human intelligence and so compete with humans would be matter of the justifiable concern.

3. Life on Mars

Our neighbouring planet Mars has been the focus of attention with regard to habitability long before the advent of powerful telescopes or space probes. With the coarsest grade of early photographic data the situation remained delightfully ambivalent in the early decades of the 20th century, with serious discussions taking place as to the possibility of intelligent life on that planet. At the turn of the 20th century Nicola Tesla claimed that he was receiving radio signals from Mars and this story had a wide airing in the press, but did not inspire scientists of the day.

With a radius of about half that of the Earth and a mass of approximately one-ninth, Mars has a surface gravity which is a little less than half of terrestrial gravity. This permits a thin atmosphere at the present time, though with not enough opacity to shield against damaging ultraviolet light at the surface. The Martian day is almost the same as the Earth day, and because the tilt of its axis of rotation is the same as of the Earth, the seasons are also similar to terrestrial seasons. On the other hand, Mars is further than the Earth from the Sun so that the Martian year is nearly twice as long as the Earth year.

Speculations about intelligent Martian life resurfaced as a result of observations of enigmatic features on the planet's surface when viewed with the low magnification telescopes available at the time. Some of these features were due to effects of Martian dust clouds that have only recently come to be fully understood; and the famous Martian canals which were assiduously mapped by many astronomers including Percival Lovell (1855-1916) were of course observational artefacts or optical illusions.

Although evidence for intelligent Martian life remained tenuous, it was difficult at the time to disprove the theory with the quality of images available at the beginning of the 20th century. An argument that could not be refuted was that if the Earth were viewed using the same telescopes and technologies from a Martian vantage point, our planet Earth would have been just as elusive over the presence or absence of intelligent life. There was no way to resolve this question unequivocally until the first Mariner probes of the 1960's sent back close-up images of the Martian surface. The answer was of course disappointingly negative, so that H.G. Wells' musings sadly came to nothing. Not only was there no evidence whatsoever of intelligent Martian life, but there were no structures that even vaguely resembled the fabled canals.

Since the earliest Mariner probes that photographed and mapped the surface at a resolution of 1km, a veritable flotilla of Mars orbiters and landers have been sent to the red planet and they continue to explore its surface and atmosphere in ever-increasing detail. The Viking orbiters launched in the 1970's quickly led to a revolution in our ideas about the composition of the red planet, and of the possibility of water on or near its surface. Vast river valleys were discovered as well as evidence of flooding in earlier epochs.

The only space mission that was explicitly directed towards searching for extant microbial life on Mars was connected with the *Viking* missions of 1976 with life detection experiments under the leadership of Gilbert V. Levin (6). These missions involved two landers carrying dedicated life detection experiments. The *Viking 1* lander touched down on 20 July 1976 on the *Chryse Planitia* near the equator; the *Viking 2* lander touched down on 3 September 1976 on *Utopia Planitia* closer to the Martian North Pole. The landers carried out biological experiments *in situ* on samples of soil, one of which was taken from beneath surface rocks. The presumption was that any microorganisms which may be present had metabolic processes similar to those of Earth microbes. The soil was treated with nutrient labelled with ^{14}C isotope, and its uptake by microbes was monitored by detecting radioactive exuded gases such as CO_2 or CH_4.

As with all innovative scientific experiments, the results turned out at first sight to be more complicated than was expected. The profusion of gas released (metabolites) that frothed out when the labelled nutrient was poured on the soil was a strong positive for life detection, but against this was the finding that the Martian soil did not show detectable amounts of even simple organic compounds. Since one could not have biology without evidence of the detritus of biology, the officialdom of NASA (Levin excluded) went public very quickly to say that the Viking results were not consistent with life. What was conspicuously missed in this assessment was the possibility that the turnover rate of life under the Mars surface conditions was so slow that the lack of organics is easily explained. Indeed, the *Viking* experiment prototype taken to the dry valleys of the Antarctic several years before the launch of *Viking* yielded nearly identical results in the presence of Antarctic microbiota. A few years later the discovery of methane in the Martian atmosphere with seasonal variations make ongoing microbiology a real prospect.

Principal Investigator Gilbert V. Levin always dissented from the official view of NASA that the Viking experiments *proved* no life on Mars. The gas released in the Viking Gas Release experiment was claimed by NASA to be more rationally explained by some inorganic chemistry involving a superoxidant, but to this day the search for the required material on the surface or in the laboratory has not been successful, and extant biology remains the most reasonable explanation of the results from Vikings 1 and 2.

In 2012 a re-examination of the data from the Viking experiments left little doubt that Levin and Straat really did discover microbial life on Mars in 1976 (6,7). Despite Levin's persistent protestations it is a sad commentary on the sociology of science that the 1976 Viking life-detection experiment, or one similar to it, was not considered desirable to include in any later mission to Mars. The Viking program ended in 1982 and another 14 years and several lost spacecraft (both US and USSR) were needed before the next successful phase of Mars exploration was resumed.

With the arrival of the *Mars Pathfinder* lander near the mouth of the Ares Valles valley on July 4 1997, further evidence of running water in the distant past was uncovered. Since then *Mars Odyssey*, *Mars Express*, and the rovers *Spirit* and *Opportunity* continue to reveal a varied terrain that may well be suited for some types of primitive extremophilic microorganisms – not dissimilar to the forms of life known to inhabit the harshest environments on the Earth.

The possibility of past life-forms more complex than microbes is also turning out to be a serious possibility with evidence gathered in the past few years, particularly in the Gale crater which most likely represents a dried-up river bed from 100's of thousands of years ago. At this time Mars seems likely to have been a hospitable life habitat; comet impacts that carried primitive life and genetic material to Earth may well have arrived there and taken root even for a short period.

The discovery of methane with a seasonal as well as diurnal variation, particularly at the Gale crater, is rather tediously interpreted as due to a geochemical process, but a biological explanation is clearly the most reasonable (8). The reluctance to conclude the existence of contemporary microbial life on Mars appears to be deep rooted in the prevailing science culture. This may well be part of the resistance to accept the wider concept of life being a truly cosmic phenomenon that takes root, evolves and flourishes whenever and wherever the right conditions prevail.

We have recently reviewed evidence that includes not only the discovery of organic molecules near the surface but also of a wide range of mineral configurations that are strongly indicative of past life and biological processes at the bottom of the Gale crater (https://www.researchgate.net/publication/365926923) (Joseph et al, 2023(9)). The tantalising possibility of multicellular life existing here on Mars prior to its occurrence on Earth now appears to be a real possibility (9). A sample of such structures shown in Fig. 1 clearly provides evidence of fossil organisms that appear to be uncannily related to similar structures known to exist in terrestrial sediments. These images have been in the public domain for some years and, in our view, have been wrongly interpreted as non-biological artefacts. Joseph et al (9) have drawn attention to a wide range of such structures (some resembling ribbed tubes) very closely resembling fossilized organisms that appear for the first time on Earth after the Cambrian explosion about half a billion years ago. The existence of similar structures on Mars at a much earlier time, possibly 3 billion years ago when rivers flowed on Mars, hints strongly at the possibility that multicellular organisms may have existed on Mars before they appeared on Earth. Some of the Gale crater "fossils" resemble terrestrial stromatolites, sponges and corals, suggesting that the later emergence of similar multicellular life-forms on our own planet were possibly derived from the asteroidal or cometary bolides carrying such lifeforms that had remained in the Earth-Mars vicinity for over 2 billion years.

Fig.1. Joseph, R.G., Rizzio, V., Gibson, C.H. et al, 2023. Fossils on Mars? A "Cambrian Explosion" and "Burgess Slale" in Gale Crater?, Journal of Astrophysics and Aerospace Technology, 11:1

Further out in the solar system the possibility of alien life still remains an open question – extending from the moons of Jupiter to larger bodies in the Kuiper Belt, planet Sedna and the still to be discovered 9th planet (10).

4. Horizontal gene transfer and the spread of intelligent alien life

Whilst amplification of microorganisms within primordial comets could supply a steady source of primitive life (archaea and bacteria) to interstellar clouds and thence to new planetary systems, the genetic products of evolved life could also be disseminated on a galaxy-wide scale (11) Our present-day solar system which is surrounded by an extended halo of some 100 billion comets (the Oort Cloud) moves around the centre of the galaxy with a period of 240My. Every 40 million years, on the average, the comet cloud becomes perturbed due to the close passage of a molecular cloud. Gravitational interaction then leads to hundreds of comets from the Oort Cloud being injected into the inner planetary system, some to collide with the Earth. Such collisions can not only cause mass extinctions of species, as one impact surely at the K/T boundary 65 million years ago, killing the dinosaurs, but they could also result in the splash-back of surface material replete with evolved genes back into space. A fraction of such Earth-debris so expelled would survive shock-heating and could be laden with viable microbial ecologies as well as genes of evolved life. Such life-bearing material could reach newly forming planetary systems in a passing molecular cloud within a few hundred million years of an ejection event.

A new planetary system could in this way become infected with terrestrial microbes' terrestrial genes that can contribute, via the well attested horizontal gene transfer process, to local biological evolution. Once life has got started and evolved on an alien planet or planets of a new system, the same process can be repeated (via comet collisions) transferring genetic material carrying the products of a local evolutionary history to other molecular clouds and other nascent planetary systems. If every life-bearing planet transfers genes in this way to more than one other planetary system (say 1.1 on the average) with a characteristic time of 40My, then the number of seeded planets after 9 billion years (lifetime of the galaxy) is $(1.1)^{9000/40} \sim 2 \times 10^9$. Such a process will undoubtedly imply the vast preponderance in the galaxy of biological systems that would include creatures possibly not dissimilar to our own with high levels of what we might describe broadly by the term "intelligence".

5. Habitable exoplanets

In 2009 NASA launched its orbiting Kepler telescope, which was specifically designed to discover planets that are the size of Earth. The detection process involved tracking down minute blinks (dimming) in the star's light when a planet transited periodically in front of it during its orbit. At present, nearly 4000 definite as well as probable detections of habitable planets have been made within only a very small sampling volume of our Milky Way (12). Most of these planets orbit red dwarf stars that are on the average twice the age of our sun. Extrapolating from the sample of present detections, the estimated total number of habitable planets in our Milky Way galaxy is reckoned to be in excess of 100 billion.

These billions of exoplanets would of course be in different stages in regard to the development of indigenous and adapted life systems, and in a fraction of such planets life may even have become altogether extinct. But with the many astrophysical processes that could operate in transferring life-bearing material across galactic distances it would now seem inevitable that such habitable planets in the galaxy would be biologically interlinked into the galactic biosphere. The processes of horizontal gene transfer that are well recognised within the context of terrestrial biology would have its widest and most natural range across the entire galaxy.

6. The Octopus

If a single discovery is to serve as a watershed in the journey to accepting our cosmic origins, it is a recent study of two related species, the squid and the octopus. The squid has an antiquity in the geological record that goes back to the great metazoan explosion of multi-celled life 540 million years ago. The octopus apparently branches out from the squid lineage about 400 million years ago, presumed to evolve from an ancestral squid. Recent DNA sequencing of the squid and octopus genomes has exploded a bombshell. The squid contains a very meagre compliment of genes adequate to serve its modest survival needs. The emergent octopus, on the other hand, has over 40,000 genes (the human has only 25,000 genes) and many of these genes apparently code for complex brain function. Others code for a highly sophisticated camouflage capability including rapid switches of colour. The octopus is incredibly more complex in structure and performance than its squid predecessor. Where did the suite of genes coding for complex brain function come from? They were not present in the ancestral squid or in any other living form that existed on the Earth at the time. The clear implication is that they came from outside the Earth – external to terrestrial biology – part of the cosmic milieu of life.

The late Arthur C. Clarke once said to one of us (Temple) that he believed the most common intelligent life forms on other planets would be octopoid. He thought two-legged land-dwellers like ourselves would be in a minority. However, octopuses as known to us on Earth are solitary creatures lacking in social groups. If octopoid beings of high intelligence on other worlds have such characteristics, they will be lacking in many social feelings which we ourselves value, and empathy would not be one of their primary characteristics. We might find that they were alien to us in more ways than one. But it is highly likely that if we survive as a species, we will indeed have to find ways of relating to octopoid beings who are not only different from ourselves but far more advanced.

7. Intelligence, Aliens and UFO's

We cannot accept that the evolution of intelligence is anywhere near the end of the road with modern humans in 2023. Rather it is more likely to be nearer the beginning. Human intelligence and human technology to enable space exploration and the creation of AI developed over less than a few hundred thousand years. Within the next million years, future developments along an upward trajectory would in our view be inevitable. These may involve further advances that involve human space travel as well as contact with ET intelligence, making the currently obsessive prevailing fear of human-engineered AI pale into utter insignificance.

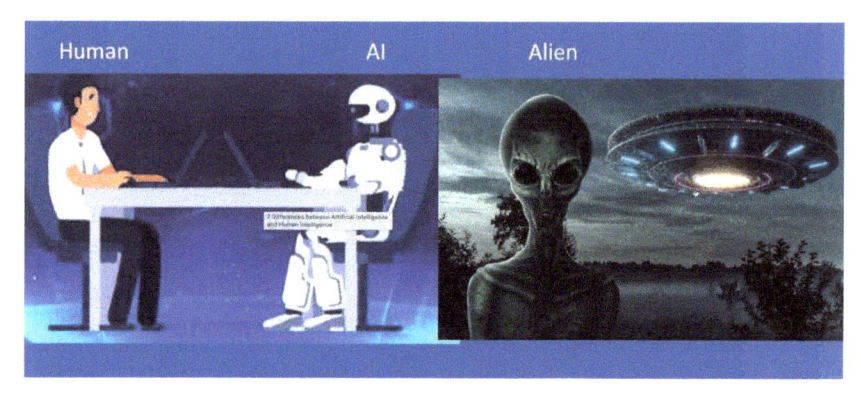

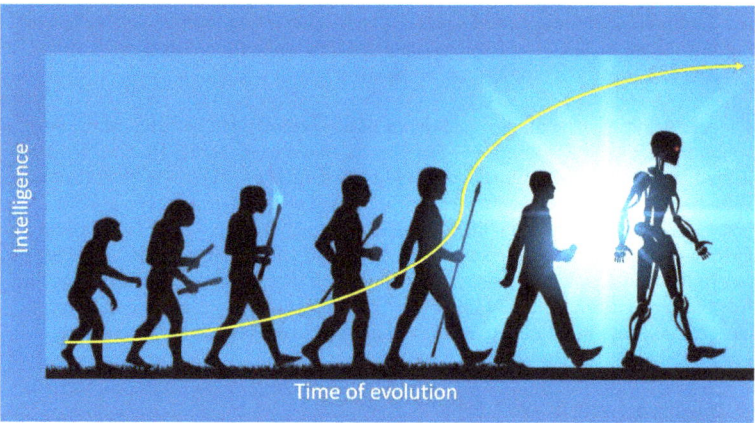

Fig.2 The evolution of human intelligence began with our ancestors over 300,000 years ago, and is on a steeply rising trajectory in 2023.

8. Inorganic intelligent entities

Two of us (Temple and Wickramasinghe) have published a paper (4) and one of us (Temple) has published a book (15) in which evidence is produced which strongly suggests that highly intelligent inorganic entities, an adjunct to organic life, may also exist throughout the Universe. Furthermore, they are life forms which do not live on planets but exist freely floating in space. They would be what are technically known as 'dusty complex plasmas', which are capable of self-organisation to such an extent that they can evolve high levels of intelligence, albeit possibly of an AI kind. Laboratory results have shown that this is indeed possible, and the many references to such findings are to be found in the two works to which we have referred. The fact that the Universe consists of 99.9% plasma rather than of 100% atoms, as had been thought in the past, suggests the possible predominance of intelligent entities in the Universe might well be "dusty complex plasmas". It could even be suggested that the Sun and all stars, which are made of plasma, as is well known, include such entities, and that they are 'alive' and 'conscious'.

There are two gigantic dusty plasma clouds between the Earth and the Moon (though not in the direct line of sight, being at the two Lagrange Points L4 and L5) which are far larger than the Earth, and which may be conscious entities. They emit no light and are extremely difficult to detect. Satellites passing through them would not be able to detect their dust particles, because their 10 nm diameters are below the level of present detectability in size. Indeed, it is possible that the entire Universe is conscious. It may be that our SETI programs should not be looking for little green men but should instead be looking for intelligent plasmas (4 and 15).

9. SETI and UFO's

In Journal of Cosmology Vol. 30 (14), we have discussed the continuing attempts at restoring SETI programmes that were started in the 1960's but which have not led to any positive results. Less respectable have been a flood of claims and counterclaims about the discoveries of UFO sightings that have not led to any decisive result. The alleged discovery of artefacts from the "Roswell incident" of July 1947 still continues to haunt us. Add to this, the ongoing reports of sightings and events reported by airline pilots and others, still remain unsolved and unresolved, and the fact that much of this data still appear to be in the domain of "classified" in both the USA and the UK.

In the present context it is relevant to note that in 2020 the United States Pentagon released a stack of formerly classified military visual recordings of UFO's (UAP's) seen from cockpit instrumentation displays of US Navy fighter jets based aboard aircraft carriers USS _Nimitz_ and USS _Theodore Roosevelt_ in 2004, 2014 and 2015. The released material also included additional footage taken by other Navy personnel in 2019.

Then, on November 3, 2021, the Department of Defence announced the establishment of its Airborne Object Identification and Management Synchronization Group (AOIMSG) as the successor to the U.S. Navy's Unidentified Aerial Phenomena Task Force, established on August 4, 2020. The AOIMSG will synchronize efforts across the Department and the broader U.S. government to detect, identify and attribute objects of interests in Special Use Airspace (SUA), and to assess and mitigate any associated threats to safety of flight and national security. All this clearly signals a growing acceptance that humans are not the only highly intelligent species even in our cosmic vicinity. Indeed, the lack of surprise in the populace at large when the recordings were recently released is the clearest confirmation of this fact.

With the vast numbers of alien habitable planets that we now know to exist (12) even within a few parsecs of the sun, it will be difficult to exclude the possibility that very much higher levels of intelligence than we have ourselves achieved in 2023 would likely to have evolved on some of these planets. So intelligent aliens must be commonplace in the Universe and the challenge is to find them! As Putoff (16) has recently discussed, a wide range of possible manifestations of alien intelligence in our midst can be identified and are fully worthy of further exploration.

References

(1) Harris M.J. et al, 2002. *Proc. SPIE Conference*, **4495**, 192
(2) Slijepcevic, P. and Wickramasinghe, C., 2021. Reconfiguring SETI in the microbial context: Panspermia as a solution to Fermi's paradox, Biosystems, **206**, 10441
(3) Temple, R., Wickramasinghe, N.C., 2019. Kordylewski Dust Clouds: Could They Be Cosmic "Superbrains"? Advances in Astrophysics 4, https://dx.doi.org/10.22606/adap.2019.44001.
(4) Wickramasinghe, N.C., Tokoro, G. abd Temple, R.. 2021. Intelligent Messages in Bacterial DNA – a Sequel to SETI? **Advances in Astrophysics, Vol. 6, No. 1,**
(5) Hoyle, F. and Wickramasinghe, N.C., 1982. *Proofs that Life is Cosmic,* Mem.Inst. Fund. Studies Sri Lanka, No. 1 (www.panspermia.org/proofslifeiscosmic.pdf)
(6) Levin GV, Straat PA, 1976. Viking Labelled Release Biology Experiment: Interim Results. Science 194: 1322-1329.
(7) Bianciardi G, Joseph MD, Ann SP, Gilbert L, 2012. Complexity Analysis of the Viking Labeled Release Experiments. IJASS 13: 14-26.
(8) Moores, JE, King, PL, Smith, CL, et al, 2019. The Methane Diurnal Variation and Microseepage Flux at Gale Crater, Mars as Constrained by the ExoMars Trace Gas Orbiter and Curiosity Observations, Geophysical Research Letters, https://doi.org/10.1029/2019GL083800
(9) Joseph, R.G., Rizzio, V., Gibson, C.H. et al, 2023. Fossils on Mars? A "Cambrian Explosion" and "Burgess Slale" in Gale Crater?. Journal of Astrophysics and Aerospace Technology, 11:1
(10) Wickramasinghe, J.T., Wickramasinghe, N.C., and Napier, W. M., Sedna's Missing Moon, *The Observatory*, **124,** 300, 2004.
(11) Wallis, M.K. and Wickramasinghe, N.C., 2004. Interstellar transfer of planetary microbiota, *Mon Not RAS,* 348, 52-61
(12) Kopparapu, R.K. 2013 A revised estimate of the occurrence rate of terrestrial planets in the habitable zones around Kepler M-dwarfs *Astrophys. J.* 767, L8
(13) Steele, E.J., Al-Mufti, S., Augustyn, K.A., et al 2018. Cause of Cambrian Explosion - Terrestrial or Cosmic? *Prog.Biophys. and Molecular Biology*, 136, 3.
(14) Wickramasinghe N.C. et al, *Journal of Cosmology* Vol. 30 references, 2023, 1-5. https://thejournalofcosmology.com/indexVol30CONTENTS.htm
(15) Temple, R., *A New Science of Heaven*, Hodder & Stoughton, London, 2022.
(16) Puthoff, H., Ultraterrestrial Models, *Journal of Cosmology,* Vol. 29, 1. https://thejournalofcosmology.com/Puthoff.pdf , 2022

A Note on a Biological Explanation for the ERE Phenomenon

Rudy Schild[1], Chandra Wickramasinghe[2,3,4] and J. H. (Cass) Forrington[5]

1. Harvard-Smithsonian Center for Astrophysics, Cambridge MA., USA
2. Buckingham Centre for Astrobiology, University of Buckingham, UK
3. Centre for Astrobiology, University of Ruhuna, Matara, Sri Lanka
4. National Institute of Fundamental Studies, Kandy, Sri Lanka
5. United States Merchant Marine Academy, Kings Point, N.Y., Cum Laude, 1972

The phenomenon of extended red emission in galactic sources, known for nearly 3 decades is best explained on the basis of biological pigments. Pigments associated with the "Red Rain of Kerala" provide a good model, although other biological pigments more generally would also serve well as a possible explanation.

Keywords: Interstellar dust, astrobiology, extended red emission

Abstract

Whilst we have attempted to account for many of the properties of interstellar dust with a primarily biological/bacterial model, one remaining set of observations that needs to be understood relates to the so-called Extended Red Emission (ERE) phenomenon. This phenomenon is also elegantly explained with a biological model of dust.

1. Introduction

The likely biological origin of infrared and ultraviolet spectral features in interstellar dust has been discussed over several decades and the relevant arguments have been reviewed in earlier papers in the present volume (1,2,3). Here we focus on perhaps one of the most bizarre spectroscopic phenomena in astronomy that defies a simple inorganic or non-biological explanation. This is the so-called extended red emission of interstellar dust (ERE), showing up as a broad fluorescence emission band over the red wavelength range 5000–7500A. The feature has been observed extensively in planetary nebulae (Furton and Witt, (4), HII regions (Perrin and Sivan, (5)), the red rectangle, and in many dark nebulae (Schild et al (6-10)). High latitude cirrus clouds in the Galaxy as well as in extragalactic systems also show the same phenomenon.

2. Role of Aromatic Molecules

We have discussed elsewhere how ensembles of aromatics can account for both UIB's (Unidentified Interstellar Bands) and the UV extinction feature 2175A in interstellar dust. The diffuse interstellar absorption bands (DIB's) in the optical spectra of stars, particularly the 4430A feature, also have possible explanations on the basis of molecules such as porphyrins (3). The ERE phenomenon has a self-consistent explanation on the basis of fluorescence of biological chromophores (pigments), e.g. chloroplasts and phytochrome. Competing models for ERE based on emission by non-biologically generated compact PAH (polyaromatic hydrocarbon) systems are unsatisfactory as is evident in Fig. 1. Hexa-peri-benzocoronene is one of a class of abiotic models that has been discussed in this context in the astronomical literature. However, the width and the central wavelength of its fluorescent emission leave much to be desired, and thus it cannot be claimed as a decisive identification of this feature.

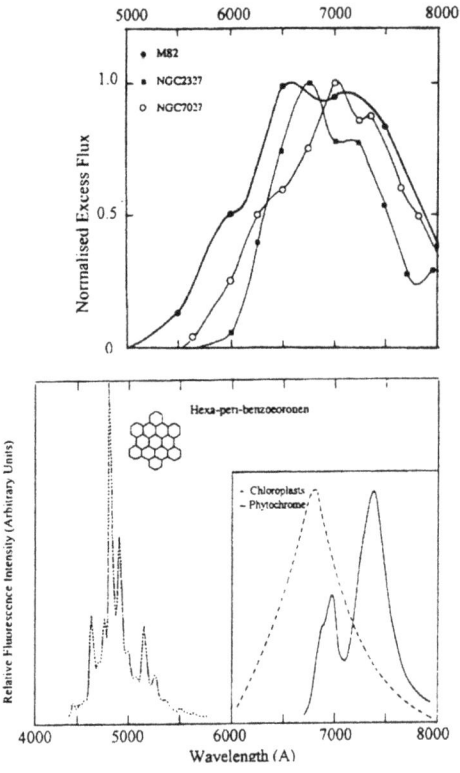

Fig. 1 The points in the top panel show the normalised excess flux over scattering continua from data of Furton and Witt (1992) and Perrin et al. (1995). The bottom right panel (inset) shows relative fluorescence intensity of spinach chloroplasts and phytochrome at a temperature of 77 K

The points in the top panel of Fig.1 show the normalized excess flux (over scattering continua) from data of Furton and Witt (1992) and Perrin et al. (1995). The bottom left panel is the fluorescence spectrum of hexa-peri-benzocoronene. The bottom right panel shows relative fluorescence intensity of spinach chloroplasts at a temperature of 77 K. The dashed curve is the relative fluorescence spectrum of phytochrome. From these comparisons it is clear that a biological explanation of the ERE phenomenon is both viable and preferable to any contrived non-biological explanation.

A recent analysis of HST observations showing red emission from the nebula NGC7023 by Witt et al (2006) have led to further uncertainties of its carrier (see Fig. 2). The standard view is that this is a nebula rich in inorganically derived PAH's (polycyclic aromatic hydrocarbons) is in serious difficulty in our view.

Fig. 2. Nebula NGC7023 showing red emission from an unidentifiable source

An alternative biological model for galactic ERE emission may be provided by studies of the red rain that fell in 2001 in Kerala, India and in 2012 in Sri Lanka. The red colour of this rain was found to be due to an unidentified microscopic organism containing a red pigment. The red rain cells in the Kerala rainfall are, incidentally, very similar to the cells recovered in the red rain of Sri Lanka some 11 years later, and the striking possibility remains that they form a component of interstellar and cometary dust.

2001 RED RAIN EVENT IN KERALA, INDIA

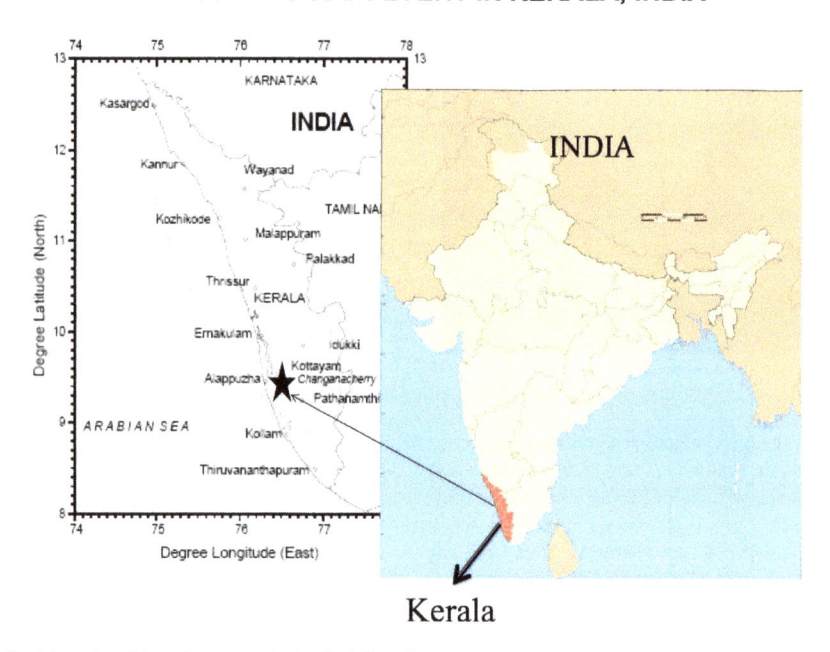

Fig. 3. Location of the red rain event in Kerala, India, 2001

Spectroscopic/fluorescence studies of the red rain have shown features that are in remarkable agreement with the ERE phenomenon (12, 13, 14). For excitation wavelengths between 412 nm and 600 nm three features show up at the wavelengths 670, 763 and 823 nm as shown in Fig.4 (12, 13).

Red rectangle emission compared with Red Rain fluorescence

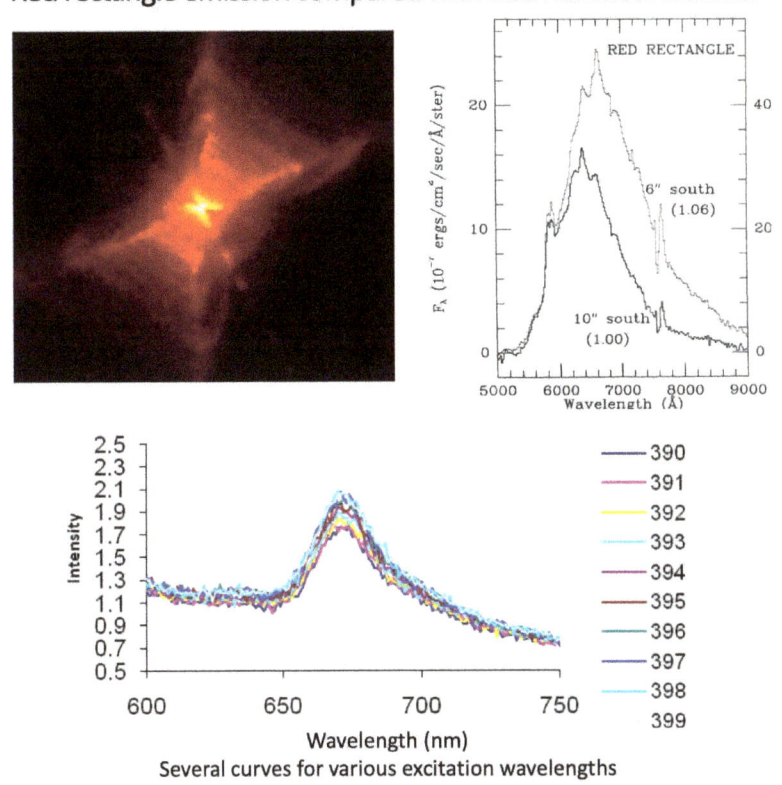

Fig.4. Emission spectra of red rain particles at different excitation wavelengths (412-600 nm) bottom curve.

We conclude by observing that this is yet another instance where the universe has its say – biology trumps over non-biology in accounting for the ERE phenomenon - the most startling manifestation perhaps being in observations of the Red Rectangle.

In this article, as well others in Vol. 30 and in earlier volumes of the *Journal of Cosmology*, the authors have clearly demonstrated that the detritus of biological life, together with pristine genetic components from which life is built, is distributed throughout the cosmos. In another article we have also argued that from the latest data obtained from the James Webb telescope, life seems to be widely distributed throughout a universe that is eternal and infinite.

In 2006 NASA scientists reported the discovery of the amino acid glycine, a fundamental building block of proteins, in the samples of comet Wild 2 returned by NASA's Stardust spacecraft. This was one of a long list of spectroscopic matches with microbial material that were discovered both in interstellar and cometary dust from 1984 to the present day.

But what of the incredible range of lifeforms that are the result of assembly of the infinity of "genetic maps" that are cosmically dispersed throughout space in the form of bacteria and viruses? Astronomical and biological evidence, in our view, continues to point decisively against the long-held concept of the spontaneous generation of life. Bacteria and viruses are an all-pervasive component of a possibly eternal cosmos; they carry new genes mostly lodged within cometary-type bodies. Such cosmic genes rain down on planets like Earth, initiating life and thereafter augmenting it to unravel a magnificent cosmic life spectacle.

The break-up of bacteria, viruses, and the detritus of life, that are dispersed throughout the vast dust clouds of space shows up in a wide range of astronomical phenomena. These include the Diffuse Interstellar Bands (DIB's), PAH's and the extended red emission (ERE) discussed in the present paper.

References

(1) Wickramasinghe, N. C., 2010. *International Journal of Astrobiology* 9 (2) : 119–129

(2) Hoyle, F. and Wickramasinghe, N.C., 1991. *The Theory of Cosmic Grains*. Dordrecht: Kluwer Academic Press.

(3) Wickramasinghe, N.C., Narlikar, J.V. and Tokoro, G.. 2023. Cosmology and the origins of life, *Journal of Cosmology*, Vol. 30, No. 1, pp. 30001 - 30013

(4) Furton, D.G. and Witt, A.N., 1990. The spatial distribution of extended red emission in the planetary nebula NGC7027, *Astropys.J.*, 364, L45-48

(5) Furton, D.G. & Witt, A.N. 1992. Extended red emission from dust in planetary nebulae, *Astrophys. J.* 386, 587-603

(6) Witt, A.N. and Schild, R.E., 1983. Hydrogenated amorphous carbon grains in reflection nebulae, *Astrophys.J.*, 325, 837-845

(7) Witt, A.N., Schild, R.E. and Kraiman, J.B., 1984. Photometric study of NGC2023 in the 3500A to 10,000A region – confirmation of a near-IR emission process in reflection nebulae *Astrophys.J.*, 281, 708-718

(8) Witt, A.N., Schild, R.E., 1985. Colors of reflection nebulae II: The excitation of extended red emission, *Astrophys.J.*, 294, 294-299

(9) Perrin, J.,-M., Sivan, J.-P., 1992. Discovery of red luminescence band in the spectrum of the Orion Nebula, *Astron. and Astrophys.*, 255, 271-280

(10) Witt, A.N., Mander, S., Sell, P.H. et al, 2008. Extended red emission in high galactic latitude interstellar clouds, *Astrophys.J.*, 679, 497

(11) Rauf, K. and Wickramasinghe, C., 2010. Evidence for biodegradation products in the interstellar medium, *Int.J.Astrobiol,* 9(1), 29-34

(12) Louis, G. and Kumar, A.S., 2006. The red rain phenomenon of Kerala and its possible extraterrestrial origin, *ApSS*, 302, 175

(13) Louis, G. and Kumar, A.S., 2013. Autoflourescence characteristics of the red rain cells, Proc. *SPIE*, 8865, 886501-1

(14) Wickramasinghe, N.C. et al, 2013. Living diatoms in the Polonnaruwa meteorite – possible link to red and yellow rain, *Journal of Cosmology*, 21(40), 9797

15) Witt, Adolf N. et al, 2006. The excitation of extended red emission: new constraints on its carrier from Hubble Space Telescope observations of NGC7023, *Astrophysical Journal*, 636, 303

Cosmicrobia: A New Designation for the Theory of Cosmic Life

Chandra Wickramasinghe[1,2,3,4]

[1]Buckingham Centre for Astrobiology, University of Buckingham, UK
[2]National Institute of Fundamental Studies, Sri Lanka
[3]Centre for Astrobiology, University of Ruhuna, Matara, Sri Lanka
[4]Institute for the Study of Panspermia and Astroeconomics, Gifu, Japan

Abstract

The common belief is that the present author and Fred Hoyle in the late 1970's embarked on a programme of work to revive the discredited, two and a half millennia-old, theory of panspermia on a whim. In this article I attempt to clear up this misconception and show that we were guided inexorably toward such a goal as a flood of new supporting data came to light from astronomy, geology as well as biology. This is an important record to set right as evidence continues to grow in the direction of supporting the theory of life being a cosmic phenomenon.

Keywords: Cosmic life, panspermia, cosmicrobia

1. Introduction

It does not require much ingenuity to observe that the prevailing theory of spontaneous generation of life on Earth lacks any compelling body of empirical fact to prove it, nor indeed support it in any significant way. Ever since Charles Darwin's *Origins of Species* was published, the much publicised "evidence" for spontaneous generation appears to be based on Darwin's letter to Joseph Hooker in 1871:

"But if (and oh what a big if) we could conceive in some warm little pond with all sorts of ammonia and phosphoric salts, light, heat, electricity etcetera present, that a protein compound was chemically formed, ready to undergo still more complex changes.... "

It was taken for granted that such a process of chemical evolution with an end directed to biology had of necessity to precede the processes of biological evolution that Darwin discussed extensively in his "*Origins of Species*" in 1959 (1). However, this does not follow logically, and the detailed processes by which it could happen in chemistry continues elude scientists.

2. Improbability argument

The operation of a living system depends upon many thousands of chemical reactions taking place within a membrane-bound cellular structure – namely a biological cell. Such reactions, determined ultimately by the order of nucleotides in DNA, are grouped into metabolic pathways that have the ability to harness chemical energy from the surrounding medium. This takes place through a series of very small steps transporting molecules into cells, building up long-chain biopolymers, and ultimately making copies of themselves. Batteries of enzymes, comprised of long chains of amino acids, play a crucial role as catalysts precisely controlling the rates of chemical reactions that ensure the proper working of the cells. In the absence of enzymes, and the exceedingly specific arrangements of amino acids within the enzymes, that are in turn coded through DNA, there could be no life (2,3).

In present-day biology, the precise "information" contained in enzymes—the arrangements of amino acids into folded chains—is transmitted by way of the coded ordering of the four nucleotides (A,T,G,C) in DNA. In a hypothetical RNA world, that some biologists think may have predated the DNA-protein world, RNA is argued to serve a dual role - as both enzyme and transmitter of genetic information. If a few such ribozymes are regarded as precursors to all life, one could attempt to make an estimate of the probability of the assembly of a simple ribozyme composed of 300 bases. This probability turns out to be 1 in 4^{300}, which is equivalent to 1 in 10^{180}, which can hardly be supposed to happen even once in the entire 13.9-billion-year history of a conventional Big Bang universe. And this is just for a single enzyme. For the totality of enzymes needed for the functioning of the simplest cell, the "Shannon" probability resources are mind-blowing to say the least. Our much publicised and criticised analogy of a tornado blowing through a junk yard leading to the assembly of a fully functioning aircraft remains as valid a comparison today as it was when Fred Hoyle first used it in a lecture to illustrate the absurdity of the proposition inherent in the theory of spontaneous generation.

3. Spreading life through the Galaxy

Besides deep-frozen cometary bodies with radioactively heated interior domains that serve as the long-term reservoirs of cosmic microbial life, freeze-dried bacteria often encased within larger particulates could be distributed on a galactic or even extragalactic scale. Introduction of such "life seeds" to planets on which life has already taken root, can also serve to spread the genetic products of the evolution of life by "horizontal gene transfer" through the Galaxy. Such a process is well attested to take place for the transport of genetic information between species of plants and animals on Earth (4) and may easily be extended to apply on a galaxy-wide scale (5).

Our own solar system, where life has existed for the past 4.2 billion years moves around the centre of the Galaxy in an orbit that is completed every 225-250 million years. In its course, gravitational interactions with nearby molecular clouds inevitably leads to perturbations of the Oort cloud of comets that surrounds our solar system causing episodes of impacts onto the Earth of life (microbe) bearing comets. In addition, dormant microbes securely preserved within clumps of interstellar dust many also be introduced from interstellar dust clouds exterior to the solar system.

The overwhelming weight of evidence favours the survival of bacteria under interstellar conditions, at any rate to the extent that makes viable interstellar transfers of microbial life between star systems inevitable. We do not require more than one in ~10^{24} iterant microorganisms to survive, until it becomes incorporated in a planet/comet forming event by which a new cycle of exponential amplification occurs. The strong feedback loop of cosmic biology depicted in **Fig.1** can account for all the astronomical data that relates to organic molecules in space.

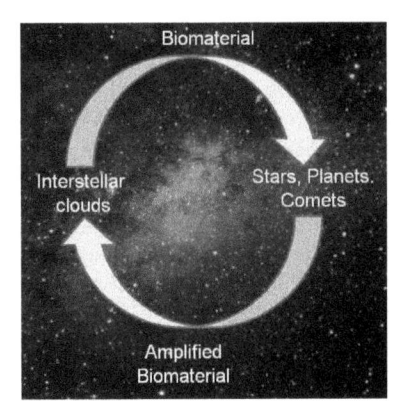

Fig.1: Bacteria and viruses expelled from a planetary system are amplified in the warm radioactively heated interiors of comets and thrown back into interstellar space, where a fraction breaks up into molecular fragments that are observed, but a non-negligible minute fraction remains viable.

The exceedingly modest requirement of microbial survival of 1 in ~10^{24} would be well-nigh impossible to violate particularly for freeze-dried microorganisms embedded within clumps of interstellar dust. This expectation has been borne out in a long series of investigations that have been conducted in the laboratory, in space, on Mars, and from the 1970's to the present day (6,7,8). In one example, the survival of colonies of *Deinococcus radiodurans* on the exterior of the international space station (ISS) for over 3 years led to surprise with a reluctant admission by the investigators that microbial life may indeed be easily transferable between distant habitats in the Galaxy.

The vast majority of bacteria in interstellar space do not and need not persist in a viable state, however. Viruses (which carry coded information in the form of DNA or RNA), compared to bacteria, have smaller target areas for cosmic-ray damage and will therefore be expected to have a relatively longer persistence in interstellar space. And both bacteria and viruses within hard-frozen interiors of comets would have an almost indefinite persistence, and for this reason comets in our theory serve as long-term cosmic reservoirs of life. Interstellar clouds which would be filled overwhelmingly with the detritus of life escaping from planetary systems as well as comets will naturally include a wide range of organic molecules as has indeed been observed over the past four decades. However, attempts to discover the roots of abiogenesis in interstellar clouds, as is currently fashionable, are manifestly futile in my view (7).

Besides the well-attested space survival properties of bacteria and viruses that are now well attested, several independent lines of evidence for the concept of life as a cosmic phenomenon have be discussed at length elsewhere (8, 9, 10, 11) and does not need further discussion. Stratospheric sampling experiments from a height of over 40 km have consistently led to positive detections of in-falling microbiota, and from samples collected in a measured volume of the stratosphere at 41km we can deduce an average in-fall rate over the whole Earth of 0.3-3 tonnes of microbes per day (12,13,14,15). This converts to some 20-200 million bacteria per square metre arriving from space every single day.

Between 2001 and the present day, this average in-fall rate of microbiota has been been amply confirmed, although it is not still widely admitted. Moreover, a crucial test to confirm their extraterrestrial origin requires access to a *Nanosims* machine to determine isotope information, and such equipment has not been made available. Recently it has also been confirmed that the exterior of the International Space Station (ISS) has deposits of bacteria which are difficult to dismiss as terrestrial contaminants (16,17). At the present time (2023) a vast body of astronomical and biological evidence continues to point to life being a cosmic phenomenon, but such evidence tends to be ignored. A paradigm shift from Earth-centred life to cosmic centred life is now long overdue, but this has been blocked persistently for reasons that are more connected with history, sociology and science politics, rather than with science itself.

4. Historical constraints and need for new name

The history of panspermia in Western philosophical tradition dates back to the pre-Socratic Greek philosopher Anaxagoras of Clazomenae (500-428 BCE). The outright rejection of panspermia by Aristotle two centuries later sealed the fate of panspermia by placing Aristotle's theory of the Spontaneous Generation of life on an almost sacred footing. Even though evidence against the validity of spontaneous generation was sporadically offered throughout succeeding centuries, a basically Aristotelean dogma has stubbornly prevailed. Every alleged "disproof" of panspermia was hailed as a triumph for orthodox scientific wisdom. The term panspermia itself, in the author's view, came to be inextricably linked to rebellion, heresy and non-conformist science. There is an urgent need to deploy a new word that encapsulates the concept of life as a cosmic phenomenon with its informational components (bacteria and viruses) distributed throughout the cosmos.

Fig.2: Fred Hoyle and Chandra Wickramasinghe at the National Institute of Fundamental Studies, Sri Lanka (Photo 1982)

When evidence for life being a cosmic phenomenon began to accumulate from the early 1980's, Fred Hoyle and the present author agonised over the question of terminology. What do well call this emerging world view? How could we communicate the essence of the new ideas to the scientific community and the wider public? Our attempts to answer such questions led us to contemplate the naming of a new emerging scientific discipline in its own right

In 1981 in the book "Space Travellers: The Bringers of Life" Hoyle and the present author (18) concluded thus:

"For a generation or more astronomers have been accustomed to thinking of star-forming episodes accompanied by the production of vast clouds of interstellar grains. The episodes are sometimes local but they are often galaxy-wide. They are thought to be triggered by some large-scale event, the after effects of which linger on for some considerable time, several hundred million years. The condensation of the exceptionally bright stars which delineate the spiral structures of galaxies has often been associated with these episodes. From our argument it seems then that even the origin of the spiral structures of galaxies may well be biological in its nature.

The potential of bacteria to increase vastly in their number is enormous. It should occasion no surprise, therefore, that bacteria are widespread throughout astronomy. Rather would it be astonishing if biological evolution had been achieved on the Earth alone, without the explosive consequences of such a miracle ever being permitted to emerge into the Universe at large. How would the Universe ever be protected from such a devasting development? This indeed be a double miracle, first of origin, and second of terrestrial confinement.

Some biologists have probably found themselves in opposition to our arguments for the proprietary reason that it seemed as if an attempt were being made to swallow up biology into astronomy. Their ranks may now be joined by those astronomers who see from these last developments that a more realistic threat is to swallow up astronomy into biology. The reality may be the birth of a new scientific disciple astrobiology."

These statements, as far as we can see represent the first documented introduction of astrobiology into the scientific arena and the start of an inevitable merger of astronomy and biology (19).

Another question that we pondered at this time related to nomenclature - the term appropriate for the concept of viable microbial life being transferred across astronomical distances. We were fully aware of the genesis of the term *Panspermia* and of his fateful history throughout the Christian era extending into modern times (Wainwright and Wickramasinghe, (20)). The frequent denials and the alleged *disproofs* of panspermia that punctated history over centuries should have signalled caution in relation to its use in the 20th century, but unwisely we adopted it perhaps *protem*, not realising the impediments that are inherent in attempting to revisit an abandoned idea. In 1997 Fred Hoyle and the present author published the book "Life on Mars? The case for a cosmic heritage?" (21) in which we made the following statement:

"The name panspermia was in our opinion ill-chosen and has probably done more to turn people off the concepts to which the name is currently applied than anything else. A better name is urgently needed. Even the crude *bugs from space* appellation is better (BFS). We might also suggest the term *Cosmicrobia* for consideration, a word that combines both cosmic and microbial meanings...."

Even belatedly I think we should now begin to use the word *Cosmicrobia* in place of *Panspermia* for the suite of ideas and results that now strongly supports the concept of life being a cosmic/cosmological phenomenon. This would be free of the stigma attached to the word *panspermia*, and might serve to herald a new beginning, thus laying a secure foundation on which future generations can build.

To conclude I will remark that signs of imminent change - a major paradigm shift from Earth centred life to life as a cosmic phenomenon - have come from many different directions and is now virtually unavoidable (22,23). The crucial data to clinch this shift must come from the results of experiments such as we have discussed here involving the recovery of microbes of space origin, and by establishing such an origin beyond a shadow of doubt. Besides the methods deployed thus far, isotope analyses of microbes recovered before they are cultured possibly deploying Nanosims equipment. In this way we would conclude that the evolution of life takes place not just within a closed biosphere on our minuscule planet Earth but extends over a vast and connected volume of the cosmos. The consequences of embracing this new paradigm will be far reaching and profound.

References

1. Darwin, C., 1857. Origin of Species, *John Murray*, London

2. Hoyle, F. and Wickramasinghe, C., 1982. Evolution from Space, *J.M. Dent*, London

3. Hoyle, F. and Wickramasinghe, N.C., 1983. Proofs that life is cosmic, *Mem.Inst.Fund.Studies, Sri Lanka*, No.1, 1983. (downloadable from *panspermia.org* website).

4. Hoyle, F. and Wickramasinghe, N.C., 2000. Astronomical Origins of Life: Steps towards Panspermia, *Kluwer Academic Press*

5. Wickramasinghe, C., 2010. The astrobiological case for our cosmic ancestry", *International Journal of Astrobiology* 9 (2) : 119–129 (2010)

6. Wallis, M.K., Wickramasinghe, N.C., 2004. Interstellar transfer of planetary microbiota, *Mon. Not. R. Astron. Soc.* 348, 52

7. Wickramasinghe, N.C., 2013. Simulation of Earth-based theory with negative results, *BioScience*, 63(2), 141

8. Wickramasinghe, D.T. and Allen, D.A., 1986. Discovery of organic grains in Comet Halley, *Nature*, 323, 44-46

9. Capaccione F, Coradini A, Filacchione G., et al 2015, The organic-rich surface of comet 67P/Churyumov-Gerasimenko as seen by VIRTIS/Rosetta, *Science* 347 (6220).

10. Levin, G.V. and Straat, P.A., 2016. The case for extant life on Mars and its possible detection in the Viking Labelled Release Experiment, *Astrobiology*, 16, 798-810

11. Wickramasinghe, N.C. 2022. Giant comet C/2014 UN271 (Bernardinelli-Bernstein) provides new evidence for cometary panspermia, *International Journal of Astronomy and Astrophysics*, 12, 1-6

12. Harris, M.J., Wickramasinghe, N.C., Lloyd, D., Narlikar, J.V., et al (2001) The detection of living cells in stratospheric samples. *Proc SPIE 4495*, 192–198.

13. Lloyd, D., Wickramasinghe,N.C., Harris, M.J., Narlikar, J.V.et al, (2002. Possible detection of extraterrestrial life in stratospheric samples. Proceedings of the International Conference entitled "Multicolour Universe" eds R.K. Manchanda and B. Paul, *Tata Institute of Fundamental Research*, Mumbai, India, pages 367ff

14. Shivaji, S., Chaturvedi, P., Suresh, K. et al, 2006. Bacillus aerius sp. nov., Bacillus aerophilus sp. nov., Bacillus stratosphericus sp. nov. and Bacillusaltitudinis sp. nov., isolated from cryogenic tubes used for collecting air samples from high altitudes, *International Journal of Systematic and Evolutionary Microbiology*, **56**, 1465

15. Shivaji, S., Chaturvedi, P., Begum, Z. et al, 2009. *Janibacter hoylei sp. nov., Bacillus isronensis sp. nov.* and *Bacillus aryabhattai* sp. nov., isolated from cryotubes used for collecting air from the upper atmosphere. *Int. J Systematics Evol. Microbiol*, 59, 2977–2986

16. Grebennikova, T.V., Syroeshkin, A.V., Shubralova, E.V. et al, 2018. The DNA of Bacteria of the World Ocean and the Earth in Cosmic Dust at the International Space Station, *Scientific*

Journal of Cosmology, Vol. 30, No. 8, pp. 30120 - 30128

World Journal 2018: 7360147. Published online 2018 April (doi: 10.1155/2018/7360147; (https://www.hindawi.com/journals/tswj/aip/7360147/).

17. Wickramasinghe, N.C. et al, 2018. Confirmation of microbial ingress from space, *Advances in Astrophysics*, 3(4), 266.

18. Hoyle, F. and Wickramasinghe, N.C., 1981. Space Travellers: the Bringers of Life, *Univ.Coll.Cardiff Press*

19. Wickramasinghe, N.C., 2002. The Beginnings of Astrobiology, *International Journal of Astrobiology*, 1 (2), 77

20. Wainwright, M. and Wickramasinghe, N.C., 2023. Life comes from space, *World Scientific*, Singapore

21. Hoyle, F. and Wickramasinghe, N.C., 1979. Life on Mars – the case for a cosmic heritage?, *Clinical Press*, Bristol

22. Steele EJ, Al-Mufti S, Augustyn KK, Chandrajith R, Coghlan JP, Coulson SG, Ghosh S, Gillman M. et al 2018 "Cause of Cambrian Explosion: Terrestrial or Cosmic?" *Prog. Biophys. Mol. Biol.* 136: 3-23, https://doi.org/10.1016/j.pbiomolbio.2018.03.004

23. Wickramasinghe, N.C., Narlikar, J.V. and Tokoro, G., 2023. New Frontiers of Astrobiology and Cosmology, *Journal of Cosmology*, Vol.30

# Search for UFOs and Aliens: Modern Evidence and Ancient Traditions

Rudy Schild[6], Chandra Wickramasinghe[1234], J. H. (Cass) Forrington[7], Robert Temple[5], Gensuke Tokoro[23], Reuben Wickramasinghe[4]

1. Buckingham Centre for Astrobiology, University of Buckingham, UK
2. Centre for Astrobiology, University of Ruhuna, Matara, Sri Lanka
3. National Institute of Fundamental Studies, Kandy, Sri Lanka
4. Institute for the Study of Panspermia and Astroeconomics, Gifu, Japan
5. History of Chinese Science and Culture Foundation, London, UK
6. Center for Astrophysics, Harvard-Smithsonian, Cambridge, MA, USA
7. United States Merchant Marine Academy, Kings Point, N. Y., Cum Laude, 1972

Abstract

An important priority for the coming decades will be to face up to the possibility that as an intelligent life-form on Earth capable of wreaking untold damage to our biosphere, we may not be alone. The possible presence of intelligent alien life in our cosmic vicinity should be faced with honesty, fortitude, as well as creativity. Recent testimony presented at a US Congressional hearing will be all too easy to dismiss as delusions or inventions, but caution needs to be exercised before such action is taken.

1. Introduction

The recent US Congress hearing on UFO reports (1) laid wide open a long-standing debate relating to such sightings (including sightings of UAP's - unidentified aerial phenomena). Three former high-ranking military officials gave evidence under oath to Congress asserting that they believe the US government knows much more about UFO's than it has already revealed. There was also astounding testimony concerning both UFO/UAP sightings and a declaration the US government is in possession of "nonhuman" biological matter. We have watched these proceedings with interest although we cannot contribute in any factual sense to the issues that have been raised including claims the US Federal government is concealing key UFO-related data from the general public. Our aim in this note is to attempt to place such claims in a wider historical and scientific context.

Although the very recent US congress inquiry into UFOs concluded with an amazingly positive result, earlier official inquiries into similar reports of sightings involved aspects of an institutional cover-up that are hard to ignore. The famous Roswell incident (2) is an iconic case in point. In July 1947 a mysterious craft evidently crashed into a site near Roswell, New Mexico, dispersing fragments of the craft material, as well as, most controversially, alleged evidence of alien bodies. The US government declared the crashed item to be a weather balloon and all the other associated evidence was discounted. We find it hard to believe those items were simply thrown in the trash. Now that the phenomenon is proven to be there, there is no longer justification for *all* governments to not release their secret files and evidence.

In diverse cultures throughout the world the idea of aliens in the sky as well as episodes of reported contact between aliens and humans are widespread. In ancient India, where the most fundamental concepts in mathematics are firmly based, including our modern number system, the concept of zero and the concept of infinity, we also find the clearest descriptions of aliens. The Vedic texts that date back at least to the 2nd millennium BCE clearly describe flying craft, alien beings and interactions between humans and aliens throughout the Vedic era. Sri Lankan mythology is also replete with similar accounts of flying machines - Ravana's Mechanical Flying Peacock. These phenomena are not considered to be feats of human engineering but imports from outer space, derived from aliens that were believed to inhabit the many worlds that are described in ancient texts, including well known Hindu, Jain, and Buddhist texts. The vivid descriptions to be found in the *Drona Parva*, which is the Seventh book of the Sanskrit epic, the *Mahābhahārata*, have long been famous for describing aerial warfare by advanced aircraft or space ships called *vimānas*. This text is at least 2000 years old, though there is no precise date for the origin of the accounts themselves. Alleged space journeys also figure heavily in Sumerian and Babylonian legends of a much greater level of antiquity, but all these are generally dismissed as religious hyperbole or figments of the fertile imagination of primitive people. The Bible also contains strange tales of Moses, Elijah, and Enoch being taken away from Earth in strange vehicles and transported to 'divine' regions of brilliant light. The mythologies of many cultures around the world contain such tales, which are a recurring theme throughout human societies in antiquity. It has been common to regard them all as fantasies. But why are there so many of these accounts, all insisting on the same things? There are too many of them. What if they are attempts by people of earlier times to describe real phenomena as best they could, in the absence of technological knowledge by the ancient authors?

2. Unexplained discoveries

There are sacred traditions still prevailing in the Dogon tribe and discussed at length by Robert Temple, which point uncannily to an alien connection (4). The modern Dogon are tucked away in a remote region of West Africa, a few hundred miles south of Timbuktu in the Republic of Mali. In 1976, Robert Temple published a book, *The Sirius Mystery*, in which he reported the Dogon claim that the Earth was once populated, a few thousand years ago, by amphibious beings from an alien planet around a star that they identified as a "companion" of the star Sirius. This claim was based on an investigation carried out by two French anthropologists, Marcel Griaule and Germaine Dieterlen who, nearly a century ago, had studied both the ancient mythology and current beliefs associated with the Dogon tribe. Temple's book was issued in a second edition, expanded by a further 50%, in 1998, after one of the Dogon assertions was unexpectedly confirmed by astrophysicists, as explained in the book; earlier editions of the book became obsolete subsequent to 1998.

Some artefacts dating to pre-industrial revolution times were also discovered in caves used by the Dogon, including one depicting a winged human like figure **Fig.2** (Temple(1)).

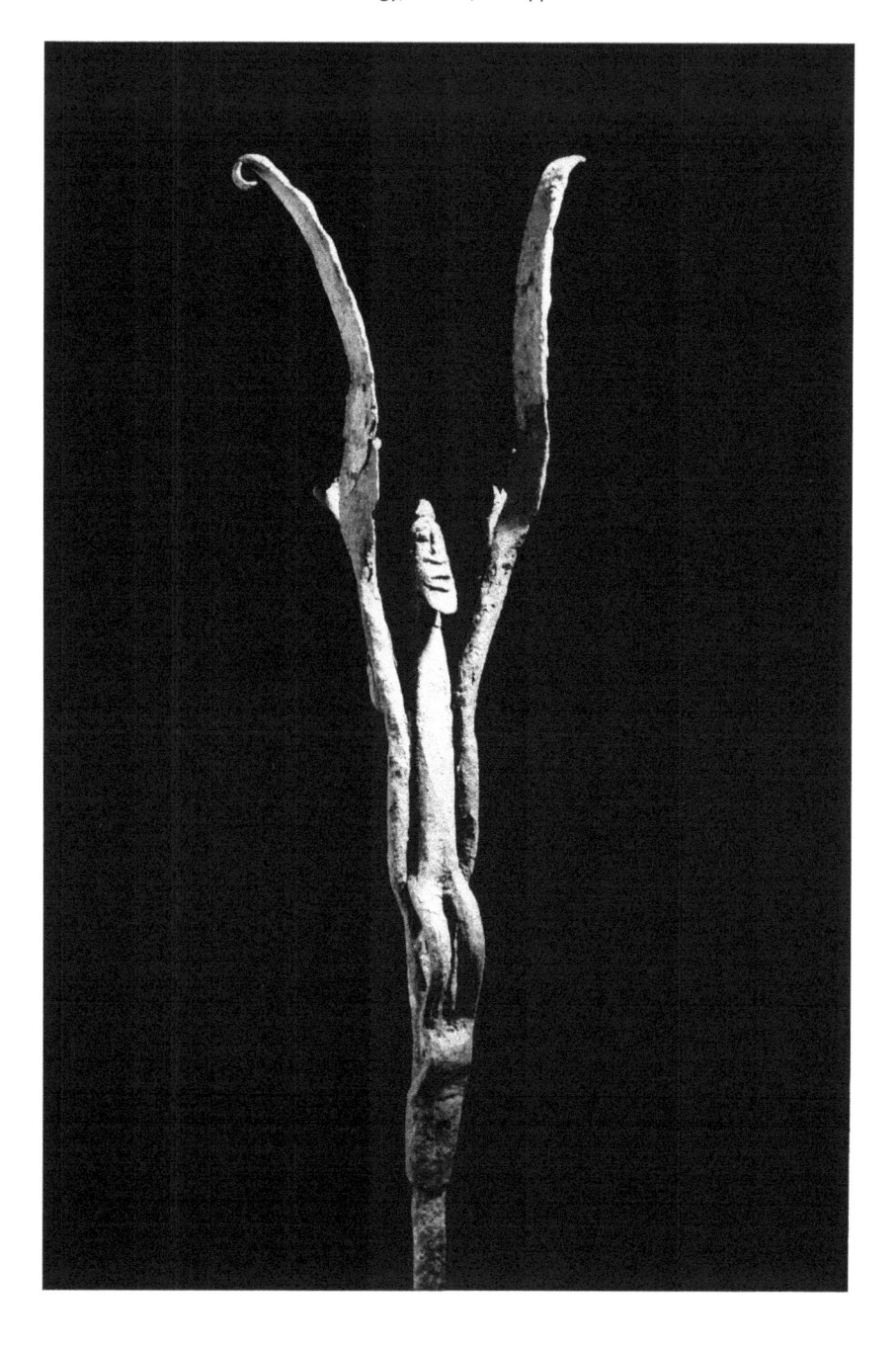

Convergence to Cosmicrobia: The Final Acceptance of Life as a Cosmic Phenomenon

Fig.2: A Three or four hundred year-old iron statue depicting a winged "alien" found in a Dogon tribe cave (1).

Some aspects of the Dogon story are similar to traditions that prevailed in ancient Egypt that included a strong focus on the star Sirius. One impressive aspect of the Dogon story is the assertion that there existed a companion star to the brightest star Sirius A - Sirius B - that was not seen or discovered until powerful telescopes were deployed in modern times. Griaule and Dieterlen also claimed that the Dogon believed that the Earth and other planetary bodies rotate on their axes and, also, that they orbit the Sun. There is also another remarkable claim that they knew that Jupiter has four moons, and that Saturn has a ring surrounding it. The Dogon also knew the difference between stars and planets, and that celestial orbits tended to be elliptical rather than circular, something which took a great deal of trouble for Johannes Kepler to discover. How did the Dogon know all of these things? They even describe the companion star to Sirius, known as Sirius B by astronomers, as being tiny, too small to see (and admit they could never see but knew it was there anyway), and that it was made of a special heavy material which does not exist on Earth. They themselves claim this information came to their ancestors in the distant past (who did not live in Mali but another land far away to the northeast) from alien visitors from a planet orbiting a third star in the Sirius system. So how did they know the difference between planets and stars? What, indeed, is the answer to what is appropriately called "the Sirius Mystery"?

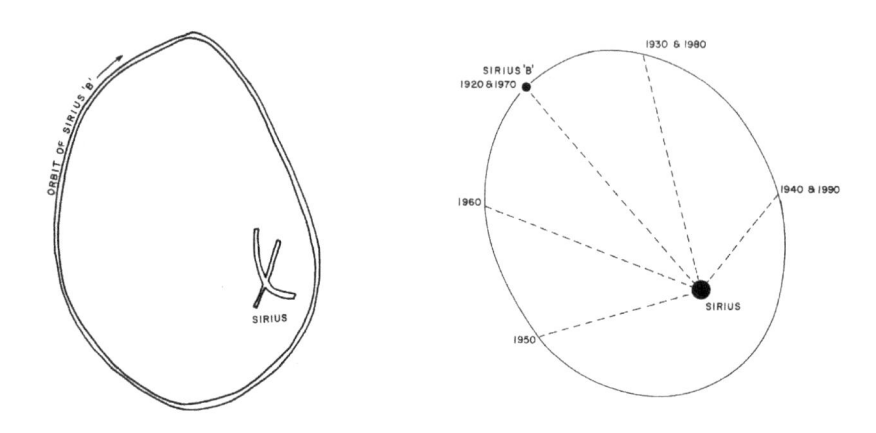

Figure 3. Left, the Dogon sand drawing of Sirius B orbiting round Sirius A. Right, the modern astronomical drawing of the same thing. How did the Dogon know about all this?

Humans seem to have wrestled with the possibility of alien life from time immemorial. Each ancient culture on record had its own particular take on what form such life might take, but they invariably converge on one aspect - aliens travelling through the skies and through space.

3. Other evidence

Besides the many astronomical sand drawings (meticulousy reproduced and published by French anthropologists) associated with the Dogon tribe depicting stars and planets, other extant ancient artefacts that are attributed in popular culture to a possible alien connection include the construction of the Egyptian pyramids, and the alignments of the three Giza pyramids on the ground (5). One of us (RT) disputes the assertion that the layout of the three main Giza pyramids represents the constellation of Orion, which he believes to be demonstrably false. Then there are also the Nazca lines (6) in Peru that are visible only from the skies; and the alignments associated with some megalithic stuctures such as Stonehenge (7) in Southern England.

There are other examples of unsolved mysteries that have taunted scientists for a long time. Some of these, like the claimed artefacts from the "Roswell Incident" may be of dubious authenticity. **We are aware of reports of alien Peruvian mummies recently revealed to the media in Mexico and we take** the point of view that it is too early for them to be critically evaluated scientifically with research on the materials and we are deeply sceptical of these mummies.

A discovery reported in 2008 involves a strange artefact embedded in sandstone at a depth of 900 metres in a coal mine in Donetsk in the Ukraine. The item that was embedded in sandstone could not be removed intact from the fragile matrix in which it was embedded. This discovery, particularly as it is located in war-torn Ukraine, has an inescapable irony, but the evidence itself looks secure enough as far as we can tell from the sketchy reports available.

Ironically, a rather similar discovery was reported by a resident of Vladivostok in far-eastern Russia (Ukraine's warring adversary) within a lump of ordinary coal that was destined to burn in a household fire. A shiny metal object was noticed protruding from a piece of friable black rock. The protrusion looked unquestionably manmade so it was brought to the attention of local scientists. The embedded object turned out to be a shiny metal bar with teeth, similar to teeth in a gear plate built so as to mesh with the teeth of a wheeled gear. The puzzle lies in the fact that the coal was formed from trees that were buried during the carboniferous era 350–300 million years ago. Occasional findings of strange objects in coal mines have been reported from time to time for more than two hundred years. These findings have tended to be dismissed out of hand, and evidence concerning them is scarce. Often these reports have only appeared as curious items in obscure local newspapers (in Germany and America), which no one chose to follow up because they were clearly "impossible". We shall never know any further details of such odd cases.

4. The Roswell Incident

Although the very recent US congressional inquiry concluded with an amazingly positive result concerning the reality of the UFO/AIP phenomenon, earlier official inquiries into similar sightings involved aspects of institutional cover-up that are hard to ignore. The now famous Roswell incident is an iconic case in point. In July 1947 a mysterious craft evidently crashed into a site near Roswell, New Mexico, dispersing fragments of the craft material, as well as, most controversially, alleged evidence of alien bodies. After many changes to this narrative coming from the relevant authorities, it was finally declared that the crashed item was a human construct – weather balloon or defence experiment either of which was well understood, and all the other associated evidence was discounted. So, the UFO phenomenon fell into serious disrepute in later years.

One of us (CW) had the good fortune of talking to Jesse Marcel Jr at a conference in the 1990's and remember the earnestness he displayed in recalling his reactions to his father's discovery of the amazing Roswell artefacts. Such reactions often tell a story that is more authentic than the record that one finds in journals and newspapers of the day.

5. Unanswered Questions

After three centuries or more of empirical science following the Enlightenment in Europe, we are still looking for answers to the question, "Are we alone as living creatures here on Earth?". Are there aliens inhabiting other planets – habitable planets that we now know certainly exist in their billions in our galaxy alone. What would intelligent aliens look like? Where should we seek to find them? One of us (RT) believes that many will be found to be octopoid, and was told by Arthur C. Clarke that it was his opinion that the majority of planet-based intelligent life forms would resemble highly intelligent octopuses.

SETI has been ongoing, albeit with changing fortunes, for over 6 decades, with no definite result. As we have recently pointed out, the presumption that our intelligent aliens would use electromagnetic signals to transmit messages to us may be an error. It is perhaps more likely that the option of microbial SETI – looking for coded messages in the silent DNA of infalling microorganisms - which is far more promising, has been overlooked (8). As we have mentioned elsewhere, our ideas of eternal cosmic life in an eternal universe, as opposed to spontaneous generation of life on our planet, leaves wide open the possibility of alien life everywhere in the galaxy and beyond.

In an eternal universe, life, itself, must also be eternal. On the other hand, if life started on Earth as a one-off against superastronomical odds, because everything was all of a sudden created at once, in less than an instant, the universe just popped into being from nowhere, and has a size and lifespan, we might be hopelessly alone.

As the James Webb space telescope has eliminated a possible age of the universe of 13.8 Gy, nullifying the Big Bang and Dark Energy/Dark Matter models of the universe, and some have just decided to double the age to 27+ Gy, either way, on the cosmic picture, we cannot expect to be at the top of the intelligence scale of life in the Universe – on the contrary we may be close to the bottom and our methods of searching for our intelligent alien neighbours might well be fundamentally flawed. In an eternal universe, we would certainly be at the bottom of the intelligence scale. Most of us on this planet don't even have toilets, sewage treatment or basic surgery, yet.

Over nearly half a century all inquiries into the existence of extraterrestrial intelligence begin by pondering the question posed by Fermi's paradox: The universe is vast and old. Therefore, we expect advanced civilizations to have developed enough by now to send their emissaries to Earth. So why have we not seen any? Fermi speculated that either such space travel was not feasible, or that our alien neighbours knew of our existence but did not consider visiting Earth was a worthwhile project. But there was another logical possibility – aliens simply did not exist, as we would understand if life developed as a miracle by spontaneous generation on Earth. Miracles don't repeat themselves!

The correct answer to, "Are there aliens?", may have been with us for decades – nay, centuries, and we were not smart enough to recognise them. Perhaps the UFO phenomenon that is currently being discussed may hold a clue.

There is another alternative. One of us (RT) has raised the suggestion many years ago that the Earth may be under a cosmic quarantine, due to the innate violence of the human species, and the inescapable fact that we are dangerous creatures to visit.

More recently, one of us (RT) has also raised the suggestion that the human species may be under sustained and continuous observation as an experiment. Humans are an amazingly creative species, but we also have a large proportion of psychopaths. The experiment may be to see whether we can 'make it'. The thin line between genius and madness could be called The Human Condition. It is possible that there is intense galactic interest in whether we can get our violence under control or whether we will destroy ourselves, so that the Human Experiment fails. It is possible that we may be one of the most interesting experiments taking place at the moment for the galactic anthropologists.

6. Outstanding scientific issues

If the accounts in the media of the recovery of biological material (non-human) is accepted as true, the question as to whether the alien biology is similar to terrestrial life in a broader sense needs to be urgently resolved. If this biology shares the same basic biochemistry as terrestrial life, then it is a startling confirmation of the ideas of panspermia and cosmic life that we have discussed in earlier papers. It may turn out to be important for medical science to analyse the details of any differences from human biology in the hope for finding remedies and cures for the many diseases that plague humanity, and the danger of any, as yet, unknown pathogens. Native Americans can sadly attest to the effects of new pathogens from other human populations, and we know pathogens jump from species to species on Earth. Could the risk of our Earthly pathogens be keeping our visitors from contacting us directly? Or, perhaps, our warlike ways...? And if we are a cosmic experiment, direct intervention in our affairs would terminate that experiment, thereby destroying the entire point of why we are interesting. It is suggested therefore that we try to continue to be interesting, if only to ensure our prolonged survival. And the best way to do that is to make civilisation work, prevent so many psychopaths rising to positions of power, and put far more effort into music, art, literature, poetry, and all the pursuits of beauty which are what we really have to offer if we are to be respected by alien civilisations.

References

1. https://oversight.house.gov/hearing/unidentified-anomalous-phenomena-implications-on-national-security-public-safety-and-government-transparency/

2. https://en.wikipedia.org/wiki/Roswell_incident

3. http://timesofindia.indiatimes.com/articleshow/38435091.cms?utm_source=contentofinterest&utm_medium=text&utm_campaign=cppst)

4. Temple, Robert K.G., 1976. *The Sirius Mystery*, Sidgwick & Jackson, London; second and greatly expanded edition: Century, London, 1998.

5. Wickramasinghe, Chandra, and Bauval, Robert, 2017. *Cosmic Womb*, Bear & Co., Rochester, Vermont, USA.

6. https://en.wikipedia.org/wiki/Nazca_Lines

7. https://en.wikipedia.org/wiki/Stonehenge

8. Slijepcivic, P. and Wickramasinghe, C. 2021. Reconfiguring SETI in the microbial context: Panspermia as a solution to Fermi's paradox, *BioSystems* 206 (2021) 10444

Life and the Universe: a Final Synthesis

N. C. Wickramasinghe[1,2,3,4], and R.C. Wickramasinghe[4]

1. Buckingham Centre for Astrobiology, University of Buckingham, UK
2. Centre for Astrobiology, University of Ruhuna, Matara, Sri Lanka
3. National Institute of Fundamental Studies, Kandy, Sri Lanka
4. Institute for the Study of Panspermia and Astroeconomics, Gifu, Japan

The long-overdue synthesis between microbiology and the universe, after many setbacks, may at long last be in sight. The full implications of the transition may well lie in the future for the benefit of the coming generations of scientists.

Keywords: cosmology, astrobiology, panspermia

1. Introduction

"Microbiology may be said to have had its beginnings in the nineteen-forties. A new world of the most astonishing complexity began then to be revealed. In retrospect I find it remarkable that microbiologists did not at once recognised that the world into which they had penetrated had of necessity to be of a cosmic order. I suspect that the cosmic quality of microbiology will seem as obvious to future generations as the Sun being the centre of our solar system seems obvious to the present generation."

Fred Hoyle in *Evolution from Space (The Omni Lecture)*, Enslow Publishers, New Jersey, 1982)

In a series of earlier papers and books evidence from a range of new discoveries were reviewed, all of which support the cosmic nature and origin of life (1-6). In this paper we take stock of the full implications that follow from these discoveries, and their possible impact on the future of science and society as a whole. We argue that at the present time in history we are at an important turning point of science, religion and philosophy, all of which appear to be inextricably interlinked.

Panspermia, as a model for the origin and distribution of life throughout the cosmos, has a very long history stretching back over 2500 years in traditions of the West. It is perhaps even older by a thousand years in traditions of the East, possibly dating back to the Vedas of ancient India or even earlier in both Egypt and China (7). The rival or competing model for life's origin – the theory of spontaneous generation - dates back to the Greek Philosopher Aristotle in the 3rd century BCE.

The name that is mainly associated with panspermia in the Western world is that of the pre-Socratic philosopher Anaxagoras, who was born in the city of Clazomenae around 500BCE, when Asia Minor was under the control of the Persian Empire. There is no doubt that Anaxagoras in his time was a towering genius, but he was also a rebel. Besides arguing the logic for panspermia – seeds of life scattered throughout the universe – his other incursions into astronomy included his claim that sun and moon were physical bodies, not deities as was widely believed; and for this heretical pronouncement he was banished from Athens and imprisoned.

Anaxagoras' arguments in relation to panspermia probably held sway in the Greco-Roman world until Aristotle came along two centuries later to argue that life *must* have originated *on the Earth* in a "Primordial Soup", by an unspecified process or processes that he called "Spontaneous Generation". Aristotle's corpus of philosophical work also included the proposition that the Earth was the centre of the universe, and it was his towering influence in maintaining this position that stalled the progress of astronomy for more than 1500 years. The Church in Rome adopted a suitably modified version of Aristotelean philosophy over a very broad front of issues, and any challenge to their authority was met with severe punishment. Galileo Galilei, who used a telescope to discover craters on the Moon as well as moons around the planet Jupiter, was famously imprisoned in 1632 for publishing his results in a book claiming that the heliocentric theory of Copernicus was correct. Giordano Bruno, poet, philosopher, astronomer, was tried by the Inquisition of Rome and was burnt at the stake in 1600 for supporting the heliocentric theory as well as maintaining that there existed other living beings on other planets orbiting distant stars. Such brutal punishments prevented dissent from Church-approved orthodoxy for a long time.

However, it is well documented that critics of spontaneous generation emerged from time to time between 1600 and 1800, as Wainwright and one of us (NCW) has recently reviewed (6). We shall see that this earlier church-enforced stricture in regard to an Earth-centred origin of life has now given way to an orthodox position that is maintained not by the church but by modern scientific institutions with equal rigour and stubbornness. Overcoming such a stumbling block in the future is the challenge we now have to face (8).

2. Louis Pasteur's challenge of spontaneous generation

Historically the first serious challenge of the long-reigning doctrine of spontaneous generation of life emerged after Louis Pasteur's classic experiments on the processes of fermentation (wine and milk) in the mid-nineteenth century. Here he demonstrated that microbial life seems always and everywhere to be derived from microbial life that had existed before (9). When a nutrient broth (in which bacteria can grow) was sealed off from the external atmosphere he showed that no bacteria will grow. It was on the basis of these experiments that Pasteur enunciated his famous dictum - *Omne vivum ex vivo*—"all life is from life that existed before".

Pasteur's experiments and his declaration turned out to be the most powerful *raison d'être* for panspermia, and his case was taken up by a number of leading contemporary physicists including, Helmholtz (10) and Lord Kelvin (11). In 1861 Lord Kelvin, in his presidential address to the British Association, stated thus:

"Hence, and because we all confidently believe that there are at present, and have been from time immemorial, many worlds of life besides our own, we must regard it as probable in the highest degree that there are countess seed-bearing meteoritic stones moving about through space. If at the present instant no life existed upon the Earth, one such stone falling upon it might, by what we blindly call natural causes lead to its being covered with vegetation."

Two decades later Nobel Prize winning Swedish Chemist Svante Arrhenius, first in a short paper in 1903 and later in his classic 1908 book "*Worlds in the Making*" (12), followed Kelvin in elaborating on the possible mechanisms by which bacteria and bacterial spores could be transferred between distant planetary systems. He pointed out that bacteria had the right sizes to be propelled by the radiation pressure exerted by sunlight or starlight and could be moved across galactic distances. It is this same mechanism that one of us (NCW) later invoked to discuss the distribution in the galaxy of interstellar dust (13). Long before the discovery of extremophiles (bacteria that can survive extreme environments) in the past half century, Svante Arrhenius had speculated that survival properties including space hardiness must exist for bacteria. He came to this conclusion from experiments that he had himself conducted by taking seeds down to near zero degrees Kelvin and finding them to survive. In this way Arrhenius introduced the concept of interstellar panspermia, a concept that was extended by one of the present authors seven decades later to become the modern theory of cometary panspermia (14).

The work of Pasteur and the powerful arguments advanced by Svante Arrhenius did not however demolish the long-held Aristotelean idea of spontaneous generation – an idea that continued to be epitomised by statements such as "fireflies emerging from a mixture of warm earth and morning dew". An abiding interest in the theory of spontaneous generation continued to persist for centuries and re-emerged with great force in the early decades of the 20th century following the work of the English Biologist John Haldane and Russian Chemist A.I. Oparin.

In the modern version of Aristotelean doctrine developed by Haldane and Oparin, an initially sterile and lifeless Earth was supposed to acquire its first life from non-living inorganic chemicals *in situ* through chemical processes taking place on our planet itself (8). This theory, even after it was later adapted to take account of inputs of organic molecules from comets and meteorites, remains fundamentally inadequate and deeply flawed for reasons that we shall discuss in this review. But first let us look at the facts relating to the history of life on the Earth, which have been unravelled mainly from studies in geology – studies searching for fossil evidence of life from the time the Earth first formed.

3. The first crucial and historic challenge of spontaneous generation

The sun and the planetary system began forming about 4.6 billion years ago by the coalescence of smaller bodies within a cloud of interstellar dust and gas known as a molecular cloud. This cloud – the proto-solar nebula as it is called - contracted under its self-gravity with the proto-Sun being formed in the hot dense centre and with the remainder of the cloud forming a "protoplanetary disc" extending out to a fraction of a light year. It is from the material within this extended cloud that planetary bodies were formed. The Earth formed from much smaller rocky bodies (asteroids) and comets colliding and coalescing shortly after the sun itself began to shine as a star.

We now know that microbial life first appeared on the Earth about 4.2 billion years ago during a period of "Late Heavy Bombardment" – period of intense bombardment by comets and asteroids. This crucial new evidence is locked away in the form of carbon spheres within larger crystals of zirconium within rocks that formed 4.1- 4.2 billion years ago (Figure 1) – and are now exposed in the *Jack Hills* outcrop in Western Australia (15). The bacterial origin of these carbon spheres is inferred from very slight departures from atmospheric values in the ratio of C12/C13 that have been found which are appropriate for their biological origin - biology showing a slight preference for C12 over C13. These bacterial fossils were clearly deposited at a time when the planet was being relentlessly bombarded by comets and meteorites and the formation of the crust of the Earth was not yet complete.

Fig. 1. Microbial carbon spherules which are inferred to be relics of bacteria within zirconium crystal from the rocks in the Jack Hills outcrop in W. Australia

It was a short while after this first injection of life that comet impacts had also delivered large quantities of water and other volatiles. This probably marked the beginnings of a habitable Earth, one that eventually became endowed with oceans and an atmosphere - a planet ideally suited for biology to take root and flourish.

4. Long periods of evolutionary stagnation – a puzzle and a clue

During an entire 2 billion-year timespan from the arrival of the first microbes 4.2 billion years ago, single-celled micro-organisms – bacteria and protozoa - were the sole life-form that inhabited our planet. At the end of this long timespan – extending over 2 billion years - we see the arrival of the first eukaryotes in the fossil record – single-celled organisms which are now endowed with a nucleus within which the cell's DNA is confined.

The first multicellular organisms appear in the fossil record as late as 700 million years before the present. Standard theories of spontaneous generation, including purely Earth-based ideas of Darwinian evolution, have great difficulty in accounting for the extraordinarily long delay in appearance between the first bacteria arriving at the Jack Hills site 4.1-4.2 billion years ago, and the first eukaryotes (cells with nuclei) a full two billion years later. Thereafter another two billion years must lapse before a veritable flood of multi-celled lifeforms including a wide variety of modern living forms suddenly appears - the so-called Cambrian explosion of life 540 million years ago.

This latter most remarkable event in the history of life on Earth marks the separation between the Pre-Cambrian and Cambrian geological epochs, and there are strong indications (from the Earth's impact history) that this event occurred during a period of *intense* comet/asteroid impacts. In our view it is such impacting bodies, bunched in time, that could have delivered new phenotypes (either intact organisms in freeze-dried condition, or new genes to be implanted in pre-existing phenotypes) – phenotypes that were hugely transformative in relation to the history of life on the Earth. The long separations between such major life-injection events on Earth are a clear indication of an astronomical phenomenon controlling the process.

5. Formidable information content of life

Proteins and enzymes, upon which the operation of all life depends, are made from folded chains of some 20 amino acids, and the precise order of these amino acids is crucial for life. Proteins are known to be coded via the precise ordering of four separate nucleotides in DNA – adenine, thymine; cytosine, guanine. After the structure of proteins and DNA came to be understood in the middle of the 20th century the mind-blowing complexity of life was revealed, and if rationality prevailed this *should* have shown the way to a cosmic view of life.

The operation of a living system depends on thousands of chemical reactions taking place within a membrane-bound cellular structure – e.g. a bacterial cell. Such reactions, determined ultimately by the order of nucleotides in DNA, are grouped into metabolic pathways that have the ability to harness chemical energy from the surrounding medium. This happens in a series of very small steps: transporting small molecules into cells, building long-chain biopolymers of various sorts, and ultimately making copies of themselves. Batteries of enzymes, composed of chains of amino acids, play a crucial role as catalysts precisely controlling the rates of chemical reactions. Without enzymes, and the specific arrangements of amino acids within the enzymes dictated by DNA, there could be no life.

In present-day biology, the precise "information" contained in enzymes—the arrangements of amino acids into folded chains—is transmitted by way of the coded ordering of the four nucleotides (A,T,G,C) in DNA. In a hypothetical RNA world, that some biologists think may have predated the DNA-protein world, RNA has been argued to serve a dual role - as both enzyme and transmitter of genetic information. If a few such ribozymes are regarded as precursors to all life, one could attempt to make an estimate of the probability of the assembly of a simple ribozyme composed of 300 bases. This probability turns out to be 1 in 4^{300}, which is equivalent to 1 in 10^{180}, which can hardly be supposed to happen even once in the entire 13.8-billion-year history of a Big-Bang universe. And this is just for a single enzyme.

A similar calculation for the precise ordering of amino acids in a minimal set of bacterial enzymes even for one of the simplest known bacteria *Mycoplasma genitalium* which has only ~ 500 genes gives an even more ridiculous probability ~1 in 10^{1000} with plausible assumptions being made (16). On the basis of such considerations, it is impossible to avoid the conclusion that the emergence of the first *evolvable* cellular life form had of necessity to be a unique event in the cosmos. If this did indeed happen on Earth for the first time, it must be regarded as a near-miraculous event and one that could not be repeated elsewhere, let alone in any laboratory simulation on the Earth. To overcome improbabilities on the scale that is involved here, it stands to common sense that we would gain immensely by going to the biggest system available—the universe as a whole.

6. Missing trunk of the tree of life?

Charles Darwin proposed that life, once it began (assumed to be on the Earth), grew like a tree from a central trunk over billions of years. The tree of life as interpreted today is divided into three distinct branches. Our own branch the eukarya, include plants, fungi and all animals including humans all comprised of cells possessing nuclei; bacteria and archaea are made up entirely of single-celled organisms lacking a nucleus, archaea being more closely related to Eukaryotes. Eukaryotes all have a distinct cell structure with organelles and a nucleus. Most of the DNA of eukaryotes are stored in their nucleus, and the cells contain a wide range of other compartments where proteins are assembled and energy is generated.

New groups of species and individual species as they emerge appear throughout the past 4.2 billion years seem to have branched off like branches of the tree of life. Sporadically we find some branches of this tree to have been pruned by extinction events, almost certainly caused by collisions, like the K/T boundary event that led to the disappearance of the dinosaurs 65million years ago.

With the introduction of new genes in other impact episodes, new shoots and new branches and have emerged like a branching tree as shown in Fig.2. This seemingly elegant analogy of a tree of life does not, however, explain any fundamental process that is involved in the origin or evolution of life. It merely offers an admittedly graphic metaphoric representation of some of the facts of terrestrial biology, facts that are indeed self-evident. To understand what really takes place, or has taken place, one has necessarily to adopt the cosmic view of life.

In our view the Earth is merely a receiving station for cosmic genes from an essentially infinite set of cosmic genes that come to be assembled over time into the magnificent panorama of life we see around us. The three kingdoms of terrestrial life – bacteria, archaea and eukaryotes – stem from an *unknown and unidentified trunk* with no discernible common ancestor, either in the genetic material of present-day life or in the fossil record. This is one more blow to theories of Earth-centred origins and evolution of life; one more way in which the Universe ultimately has its say, ultimately forcing the correct point of view to emerge.

THE TREE OF LIFE

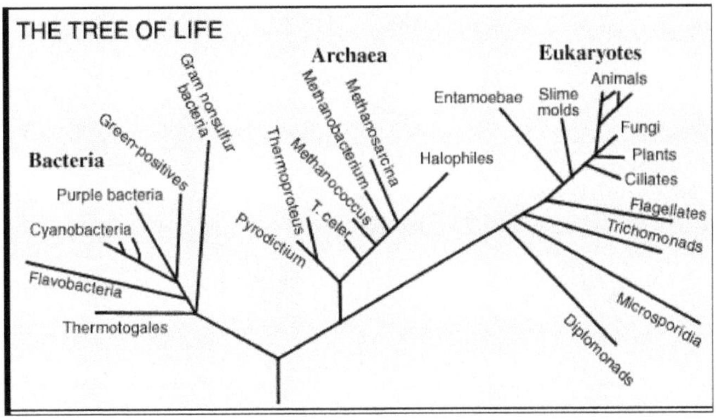

Fig.2. The schematic tree of life

7. Growing astronomical evidence for panspermia

When the senior author (NCW) first began to explore the nature of interstellar dust in 1960 the vast amounts of dust that exist in the Milky Way were thought to be comprised of micrometre sizes ice particles, similar to the ice grains found in the cumulous clouds in the Earth's atmosphere (13). Work carried out in the 1960's eventually showed that this model of cosmic dust was wrong, and instead the emerging idea was that the cosmic dust was made up mainly of the element carbon. The form of carbon in cosmic dust was next studied by examining the light from distant stars as it traversed the dense clouds in interstellar space, leaving signatures revealing the chemical make-up of the dust.

By the mid-1970's there was clear astronomical evidence for the widespread occurrence in the galaxy of organic molecules (17). At about this time Infrared Astronomy was "born" and infrared observations of interstellar dust were beginning to show spectral features in the mid- and near-infrared wavelength region which could not be reconciled with a combination of inorganic silicates and water-ice as was previously believed.

This exceedingly modest requirement of microbial survival would be impossible to violate particularly for freeze-dried microorganisms embedded within particles of interstellar dust. The vast majority of bacteria in interstellar space does not, and need not, persist in viable state. Interstellar clouds could be filled overwhelmingly with the detritus of life which takes the form of genetic fragments that could include viruses and viroids, as well as a wide range of organic molecules.

8. Direct spectroscopic proof

By the early 1980's we had accumulated enough evidence to claim that the chemical make-up of cosmic dust judged by the way they absorbed light was uncannily similar to bacteria and viruses as for instance shown in the comparison displayed in the left-hand panel of Fig.3 (18).

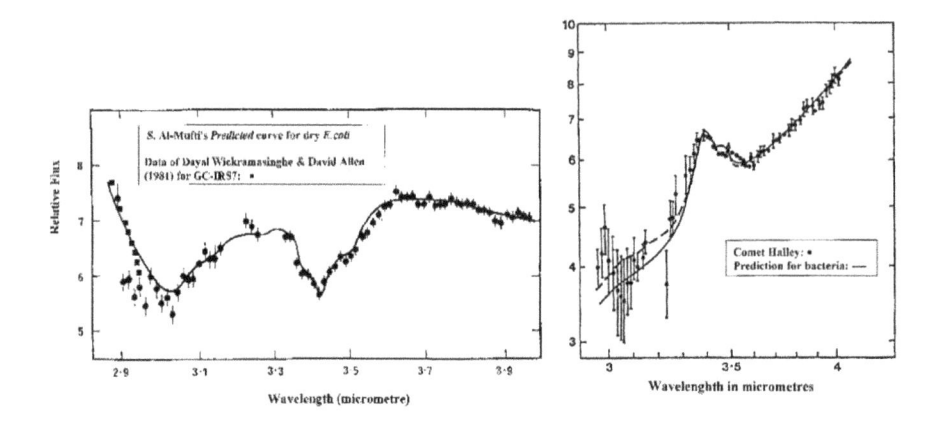

Fig. 3 Left panel: Comparison of the normalized infrared flux from GC-IRS7 (ref.18) with the laboratory spectrum of *E coli*. Right panel: Emission by dust coma of Comet Halley observed by D.T.Wickramasinghe and D.A. Allen on March 31, 1986 (points) compared with normalized fluxes for desiccated E-coli at an emission temperature of 320K. The solid curve is for unirradiated bacteria; the dashed curve is for X-ray irradiated bacteria.

Today, a wide range of astronomical observations of a similar kind all point to the widespread existence of cosmic dust with a composition resembling that of living material (see review in Wickramasinghe (19)). The prevailing reluctance, however, is to admit that that the distribution of organic molecules reflected in Fig.3 are really the products of biology. The fashionable and conventional point of view nowadays is to assert without any proof that organic chemistry is occurring everywhere, and the resulting chemicals happen perchance to match exactly the spectral behaviour of desiccated bacteria! Furthermore, it is maintained against all the odds that terrestrial life originated in a geological instant *in situ* on the Earth, after organic molecules from space came to be delivered possibly by the agency of comets.

More recent studies of other comets have yielded generally similar results. The European Space Agency's Rosetta Mission to comet 67P/C-G has provided the most detailed observations that satisfy all the consistency checks for biology and the theory of cometary panspermia (20).

A further prediction of cometary panspermia is that microbial life must be transferred beyond Earth to other planetary bodies in the solar system. Mars, Venus, and the Jovian moon Europa are a sample of the "nearby" places that are by no means written off in regard to the possibility of extant microbial life (5). We are confident that future space exploration in the coming decades will undoubtedly result in discovering long-awaited positive results for the widespread occurrence of microbial life in solar system.

9. Habitable exoplanets and transfer of genes on a galaxy-wide scale

Figure 4: Artist's impression of a habitable planet orbiting a red dwarf star Proxima Centauri (Courtesy NASA).

We must next ask the question: outside of the confines of our solar system, where is the evidence that other suitable homes for life exist? In 2009 NASA launched its orbiting Kepler telescope, which was specifically designed to discover planets that are the size of Earth. The detection process involved tracking down minute blinks (dimming) in the star's light when a planet transited periodically in front of it during its orbit. At the present time nearly 4000 definite as well as probable detections of habitable planets have been made within only a very small sampling volume of our Milky Way (21). Most of these planets orbit red dwarf stars that are on the average twice the age of our sun. Extrapolating from the sample of detections in our local vicinity the estimated total number of habitable planets in the entire Milky Way galaxy is reckoned to be in excess of 100 billion. *Proxima Centauri b* (also called *Proxima b*) is closest habitable exoplanet orbiting the red dwarf star Alpha Centauri at a distance of some 4.2 light years from the Sun. (Fig.4).

Whilst comets could supply a source of primitive life (bacteria, viruses and genes) to interstellar clouds, and thence to new planetary systems and embryonic exoplanets, the genetic products of evolved life (local evolution) could also be disseminated on a galaxy-wide scale (22). At the present time our solar system, which is surrounded by an extended halo of some 100 billion comets (the Oort Cloud) replete with microbial content as we have seen, moves around the centre of the galaxy with a period of 230 My (See Fig.5).

Every 40 million years, on the average, the cloud of comets in our solar system becomes gravitationally perturbed due to the close passage of a giant molecular cloud. Such gravitational interactions lead to hundreds of comets from the Oort Cloud being thrown into the inner regions of our planetary system, some to collide with the Earth. Such collisions do not only cause extinctions of species (as one impact surely did 65 million years ago, killing the dinosaurs), but they could also result in the expulsion of surface material containing viable bacteria and spores back into deep space.

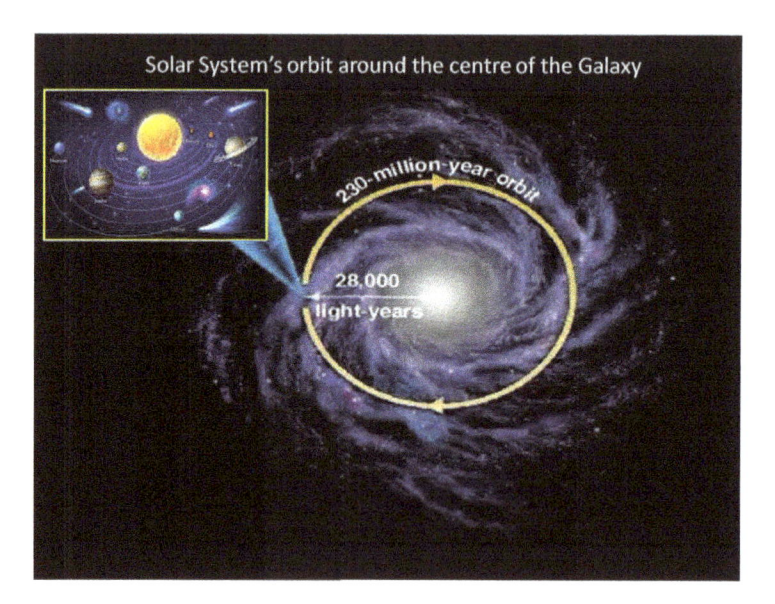

Fig.5: Path of the solar system around the centre of the Galaxy

A mechanism can thus be identified for the genes of evolved Earth-life to be transferred to alien habitable exoplanets. A fraction of the Earth-debris so expelled survives shock-heating and could be laden with viable microbial ecologies as well as the genes of evolved life. Such life-bearing material from the Earth could reach newly forming planetary systems in the passing molecular cloud within a million years of the ejection event. A habitable exoplanet could then become infected with terrestrial microorganisms and terrestrial genes that can contribute to the process of local biological evolution on a distant exoplanet. Once life has got started and evolved on an alien planet or planets of a new system, the same process can be repeated (via comet collisions) transferring a new compliment of genetic material carrying local evolutionary 'experience' to other molecular clouds and other nascent planetary systems.

If every life-bearing planet transfers viable genes in this way to more than one other planetary system (say 1.1 on the average), with a characteristic time of 40My then the number of seeded planets after 9 billion years (lifetime of the galaxy) is $(1.1)^{9000/40} \sim 2 \times 10^9$. Such a large number of 'infected' planets illustrates that Darwinian evolution, involving horizontal gene transfers, must operate not only on the Earth or within the confines of our solar system, but on a truly galactic scale. Life throughout the galaxy on this picture would inevitably constitute a single connected cosmic biosphere.

10. Cosmic bacteria entering Earth's stratosphere

One crucial test of the theory of cometary panspermia is to probe the stratosphere for currently in-falling alien genetic systems – bacteria and viruses. The first such dedicated effort to test the idea of bacterial in-fall from comets was made in 2001 by a group of UK scientists in collaboration with scientists at ISRO (Indian Space Research Organisation). Unequivocally positive detections of in-falling microbiota were made, and the number of bacterial cells collected in a measured volume of the stratosphere at 41km led to an estimate of a total in-fall rate over the whole Earth of 0.3-3 tonnes of microbes per day (23). This converts to some 20-200 million bacteria *per square metre* arriving from space every single day!

Bacteria have also been recovered more recently from the exterior of the international space station which orbits at a height of 400 km above the Earth's surface (24). More expensive and sophisticated tests need to be carried out even on the samples collected so far, if we are to prove beyond a shadow of doubt that these microbes are unequivocally alien. One such test involves the deployment of a rather rare laboratory resource – a *NanoSIMS* machine. This will determine the isotopic composition of carbon, oxygen and other constituent elements within the individual bacterial cells, and if the composition turns out to be non-terrestrial, it is QED! We have finally won, and our opponents can go home! The tight control of the relevant experimental resources worldwide has so far prevented access to this equipment.

In the absence of such experiments being done sceptic is thus left in a seemingly comfortable position to assert, if he so wished, that what we have found in our balloon samples in 2001 and 2009 were terrestrial contaminants (23,24). The situation we have described is just one instance of a totalitarian control of science that is hindering progress.

11. Relevance of new developments

Recent observations using the James Webb Space Telescope have challenged the conventional model of the singular Big-Bang cosmological model in which the universe originated some 13.8 billion years ago, giving it an age close to three times the age of the Earth (5). Galaxies observed at very high redshifts have pushed this age estimate to >> 13.8 billion years, supporting alternative models of the Universe which have an open timescale (Hoyle et al (25)). The beginnings of life in such a universe would imply the genetic components of life (spermata) are probably eternally present, so a discrete moment of origin becomes irrelevant.

It is obvious that much work still remains to be done to understand the evolution of the Universe at large. So far as the Earth is concerned the picture emerging is in a sense creationary - fully created genetic components relevant for all evolutionary contingencies were already at hand available from an external eternal universe. It might be argued that this is not really an explanation in any very deep sense of the ultimate origin and evolution of life such as orthodox biology incorrectly claims to offer. Rather it is a logical and rational *rejection* of those claims that form an impediment to gaining any deep understanding of the real nature of life and its inexorable cosmic heritage. It is a gateway leading to a different landscape with new vistas beckoning that will be the privilege of future generations to explore.

We have pointed out earlier that non-scientific constraints have all too often stood in the way of exploring new conceptual landscapes. For nearly three quarters of the 20th century during the *first* Copernican revolution enormous difficulties were experienced in getting the simplest of facts accepted when they ran counter to established religious belief. Science was of course nurtured within cultures that were predominantly Judeo-Christian and they accordingly championed a world view that was both Earth-centred as well as human-centred. The subsequent acceptance of a heliocentric world view, and later of Darwinian evolutionary thinking, already represented uneasy compromises. As we have discussed in this article the further extension of the Copernican revolution to remove Earth from the centre of biology is meeting a similar hostility as we have discovered over the past few decades.

In the 13th century CE Thomas Aquinas had integrated Aristotelean philosophy into Christian theology and this gave added weight to an Earth-centred Ptolemaic world view. In this emergent world view the Earth was both the physical and also the biological centre of the universe. As we have stated earlier the Copernican revolution of the 16th century eventually removed the Earth from its position of physical centrality in the Universe, but biology continued to remain firmly Earth-centred with spontaneous generation holding sway almost to the present day.

It is, however, a fact that science in the modern world is no longer the monopoly of Christendom nor of Judeo-Christian philosophy. At the present time we are witnessing the rapid emergence of scientific and technological cultures that have non-Christian roots, as for instance in India, Japan and China. As these new scientific cultures expand and grow in influence, we can wonder what their effect on the world scene would be. It is worth noting that Buddhism, which is an important cultural force in Japan, China, India and Sri Lanka, maintains a refreshingly open attitude and is relatively free of dogma. Siddharth Gautama Buddha, the founder of Buddhism, stressed the importance of discovering truth for oneself. On his deathbed he instructed his chief disciple Ananda thus:

"You should live as lamps unto yourselves. Hold fast to the lamp of Truth. Take refuge only in Truth. Look not for refuge to anyone beside yourself.... And those who now in my time or afterwards live thus, they will reach greatness if they are desirous of knowledge." – *Mahaparinibbana sutra No.16*). This is of course exemplary advice to scientists even in the present day.

Buddha's own vision of the world was also remarkably post-Copernican, even in the 5th century BCE. He described a Universe comprised of billions of "Minor World Systems" each resembling our own solar system. In the oldest Buddhist texts, which are in the form of dialogues with his disciples, it is stated that in the infinite space of the Universe there exists "billions of suns, billions of moons, billions of Jambudhipas, billions of Aparagoyanas, billions of Uttarakurus, billions of Pubbavidehas...." Jambudhipa and the other names are words to describe the inhabited regions of the Earth known to the people living in North India at the time. Throughout the extensive dialogues of the Buddha, it is amply clear that the Buddha viewed life and consciousness (which he thought was associated with all life) as cosmic phenomena, linked inextricably to the structure of the universe as a whole.

It is thus clear that in many important respects the traditions of Buddhism are well suited to extending the Copernican revolution to its next phase – accepting life as a cosmic phenomenon. If such traditions prevail astronomy and biology may at last be freed of its medieval fetters.

References

(1) Wickramasinghe, N.C., Narlikar, J.V. and Tokoro, G., 2023. Cosmology and the Origins of Life, *Journal of Cosmology*, Vol. 30, No. 1, pp. 30001 - 30013

(2) Wickramasinghe, N.C., 2023. Life beyond the limits of our planetary system, *Journal of Cosmology*, Vol. 30, No. 1, pp. 30020 - 30024

(3) Wickramasinghe, N.C. and Tokoro, G., 2023. Quest for life on Jupiter and its moons, *Journal of Cosmology*, Vol. 30, No. 3, pp. 30030 - 30034

(4) Wickramasinghe, C., Tokoro, G., Temple, R. and Schild, R., 2023. Reluctance to admit we are not alone as an intelligent lifeform in the cosmos, *Journal of Cosmology*, Vol. 30, No. 4, pp. 30040 - 30053

(5) Wickramasinghe, C., Schild, R. and Forrington, J.H., 2023. Second Copernican Revolution, *Journal of Cosmology*, Vol. 30, No. 5, pp. 30060 - 30071

(6) Wainwright, M and Wickramasinghe, N.C., 2023. Life comes from space – The decisive evidence, *World Scientific Publishers*, Singapore

(7) Temple, R., 2007. The history of panspermia: astrophysical or metaphysical, *International Journal of Astrobiology*, 62, 169-180

(8) Wickramasinghe, C., Wickramasinghe, K. and Tokoro, G., 2019. Our cosmic ancestry in the stars, *Bear & Co.*, Rochester, USA

(9) Pasteur, L., 1857. *C.R.Acad.Sci.*, 45, 913-916, 1857

(10) von Helmholtz, H., 1874. In *Handbuch de Theortetische Physik*, (eds W. Thomson and P.G. Tait Vo1 (Part 2) Brancscheig

(11) Thomson, W., 1871. *British Association for the Advancement of Science*, Presidential Address

(12) Arrhenius, S., 1908. Worlds in the Making, *Harper*, London

(13) Wickramasinghe, C., 1967. Interstellar Grains, *Chapman and Hall*, London

(14) Wickramasinghe, C. (ed.), 2015. Vindication of Cosmic Biology, *World Scientific Press*, Singapore

(15) Bell, E.A., Boehnke, P., Harrison, T. et al, 2015. Potentially biogenic carbon preserved in a 4.1 billion-year-old zircon, *PNAS*, 112 (47) 14518-14521
www.pnas.org/cgi/doi/10.1073/pnas.1517557112

(16) Hoyle, F. and Wickramasinghe, N.C., 1982. Evolution from Space, *J.M. Dent*, London

(17) Wickramasinghe, N.C., 1974. Polyoxymethylene polymers as interstellar grains, *Nature* 252, 462–463.

(18) Hoyle, F., Wickramasinghe, N.C., Al-Mufti,S., Olavesen, A.H. and Wickramasinghe, D.T.,1982. Infrared spectroscopy over the 2.9-3.9 micron waveband in biochemistry and astronomy, *Astrophys. Sp.Sci.*, 83, 405

(19) Wickramasinghe, C., 2010. The astrobiological case for our cosmic ancestry, *Int.J.Astrobiol.*, 9(2), 119.

(20) Wickramasinghe, N.C., Wainwright, M., Smith, W.E., Tokoro, G., Al Mufti, S. and Wallis, M.K., 2015. Rosetta Studies of Comet 67P/Churyumov–Gerasimenko: Prospects for Establishing Cometary Biology, *J.Astrobiol Outreach*, 3:1

(21) Kopparapu, R.K. et al., 2013. Habitable zones around Main-Sequence stars: New estimates, *Astrophys.J.* 765, 131.

(22) Wallis, M.K. and Wickramasinghe, N.C., 2004. Interstellar transfer of microbiota, *Mon.Not. R.A.S*, 384(1), 52-61

(23) Harris, M. J., Wickramasinghe, N. C., Lloyd, D., *et al.*, 2002. Detection of living cells in stratospheric samples. *Proc. SPIE*. 4495, 192–198. doi: 10.1117/12.454758

(24) Wickramasinghe, N.C., Rycroft, M.J., Wickramasinghe, D.T., *et al*, 2018. Confirmation of Microbial Ingress from Space, *Ad.Ap,* 3(4), 206

(25) Hoyle,F., Burbidge,G. and Narlikar,J.V., 2008. A different approach to cosmology, *Cambridge University Press*

STANDARD BIG-BANG COSMOLOGY FACES INSURMOUNTABLE OBSTACLES?

Jayant V. Narlikar[5] and N.Chandra Wickramasinghe[1,2,3,4]

1.Buckingham Centre for Astrobiology, University of Buckingham, UK
2.Centre for Astrobiology, University of Ruhuna, Matara, Sri Lanka
3.National Institute of Fundamental Studies, Kandy, Sri Lanka
4.Institute for the Study of Panspermia and Astroeconomics, Gifu, Japan
5.Inter-University Centre for Astronomy and Astrophysics, Pune 411 007, India.

Abstract
A wide range of new observations, including the most recent data from the James Web Space Telescope (JWST), may be on the verge of disproving the long- held conventional model of cosmology – a Big-Bang Universe.

Keywords: *Big-Bang theory, High-Redshift galaxies, JWST data, Conflict with data, Alternative cosmologies*

1. Introduction

A relatively simple model of the Universe proposed by the Roman Catholic priest/physicist George Lamaitre in 1927 posited that the Universe originated *ex nihilo* from a state of ultra-high density at a definite moment in time. Scarcely two years later the discovery by Edwin Hubble of the redshift of spectral lines in galaxies led to the currently accepted model of the expanding Universe – the so-called class of Big Bang cosmologies (Hubble, 1929; Narlikar, 2010; Hoyle et al, 2000). At the time the redshifts of the spectral lines of galaxies were solely attributed to the recessional velocities of galaxies and thus an elegant and simple model of the expanding universe was born. The origin itself remained shrouded in mystery from the outset, with a subsequent high density - high temperature phase of the early universe gradually acquiring the garb of physics.

From this time on the rapid development of more powerful telescopes and new technologies, including space telescopes, has provided an ever-increasing body of data that needs to be accommodated within the framework of viable cosmological models. The requirement of such models is that they must be consistent with the abundances and distribution of light elements as well as features of the cosmic microwave background later came to be discovered in 1964.

The Big-Bang cosmological model that was initially simple has, with the advent of more data and subsequent refinements, turned out to be complex and riddled with inconsistencies as was pointed out by Hoyle, et al (2000). Current "standard" Big Bang cosmological model involves the addition of dark energy comprising 69% of the mass energy density of the universe, dark matter comprising 25%, leaving "ordinary" atomic matter to make up just 5%. Current cosmological modelling is beginning to look more and more like politics than a quest to understand the true nature of the world. There is a tendency to seek approbation by learned societies and academies above all else.

2. Bare facts about the Universe

What do we really know about our universe?

According to Big Bang theory, the universe was born out of a cosmic explosion 13.8 billion years ago, the universe rapidly inflated and then cooled, It is still expanding at an increasing rate and mostly made up of unknown dark matter and dark energy. To account for new observations the 'standard model' is further enriched with new assumptions.

This well-known story is usually taken as a self-evident scientific fact, despite the relative lack of empirical evidence—and despite a steady crop of discrepancies arising with observations of the distant universe.

In recent months, new measurements of the Hubble constant, the rate of universal expansion, suggested departures between two independent methods of calculation. Discrepancies on the expansion rate have huge implications not simply for calculation but for the validity of the currently accepted standard model. Other discrepancies with the standard Big-Bang model have also been exposed by one of us (JVN), Fred Hoyle, G. Burbidge and H.C. Arp amongst others over several decades (Arp et al, 1990; Narlikar, 2002;

Another class of Universe models of infinite age, that are not unlike the class of quasi-steady state models developed by Narlikar and Hoyle (*see* Hoyle et al, 2000; Narlikar, 2002), was proposed and developed by Roger Penrose and his colleagues (Penrose, R. 2010).

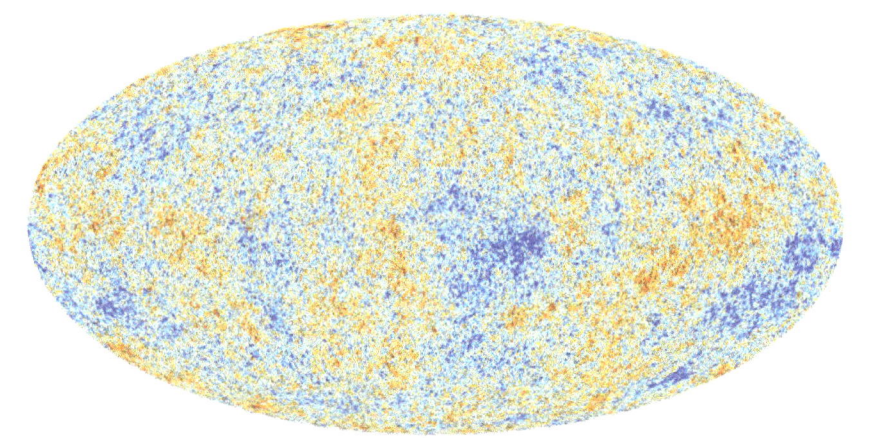

The main evidence cited for this theory is the appearance of unexpected hot spots in the cosmic microwave background (CMB) that are posited to have been produced by black holes that evaporated before the 13.8 by Big Bang-age. This is the main basis of the claim by Penrose and his collaborators including Steinhardt and Turok (Penrose, 2010) in support of a oscillating universe model wherein our present universe is one phase in a potentially infinite cycle of cosmic extinctions and rebirth.

The overall logic of the widely accepted standard model of a universe originating in a cosmic explosion 13.8 billion years ago appears manifestly in tune with Judeo-Christian ideas of creation, and it cannot be denied that this circumstance conferred upon it an extra dimension of "plausibility". Following the initial ultra-hot creation event, the universe is then posited to have undergone an epoch of "inflation", cooled and thence continued to expand at an accelerating rate. All this is tacitly taken as being firmly established fact although this was of course not true.

3. Early history of the Big-Bang Universe

A rough sketch of the early history of the universe in this standard cosmological model may be summarised thus:

- **One second post Big-Bang** - The universe is made up of fundamental particles including quarks, electrons, photons and neutrinos. Thereafter the universe is supposed to have continued to expand, though not as rapidly as during inflation, and cooled in the process.

- **Earliest post Big-Bang epoch** - As the universe continued to cool the four fundamental forces of nature i– gravity, the strong force, the weak force and the electromagnetic force - is posited to have made their debut. Neutrons and Protons begin to form when the temperature of the universe was some 10^{37} Kelvin or hotter.

- **3 minutes after the Big Bang** - Protons and neutrons began to combine to form the nuclei of simple elements. The temperature of the universe was still incredibly high at about 10^9 Kelvin.

- **400,000 years after the Big Bang** - The universe had cooled to about 3000 K and thereafter electrons combine with hydrogen form helium nuclei. Nucleosynthesis begins.

- **Some 300 million years after the Bang** – The first stars begin to form.

4. Conflicts with observational data

Perhaps the most serious challenge to conventional cosmological models has come from recent observations using the James Webb Space Telescope discovered (Bradley et al, 2023). Galaxies with ever-increasing redshifts values are being discovered and the current well-established Galaxies with ever increasing redshifts are being reported from JWST observations. The record to date is for the galaxy JADES-GS-z14-0 which appears to be the most distant confirmed galaxy with a redshift of 14. This is reckoned have formed 250 million years post-Big-Bang, which already stretched the standard model beyond the limit.

One of the main criteria to support such Big-Bang cosmological models involves measurements of the so-called Hubble constant, the rate of expansion of the universe. Recent studies have exposed major discrepancies between two independent methods of calculation, and this is beginning to cause doubts in the minds of cosmologists. Such discrepancies in the value of the Hubble constant has a relevance not only in estimating the distances of galaxies, but for the validity of the entire so-called "standard cosmological model" including the hypotheses of dark energy, dark matter and inflation.

Alternative cosmological models that are free of these discrepancies include the Quasi-Steady-State cosmologies of Fred Hoyle and one of us (Narlikar) (Hoyle, et al 2000; Narlikar, 2002) as well as the oscillating universe models proposed by Steinhardt and Turok and Penrose (Penrose, 2010). In both cases we have a class of models involving a universe that are infinite in time and space, and which are better able to explain many aspects of recent observations.

The recent discovery of a cluster of galaxies with excessively high redshifts creates problems. If their redshifts are true cosmological redshifts the age of the universe would far exceed the canonical Big-Bang age of 13.8 billion years. However, the claim is that much of the redshift could arise from non-cosmological effects (eg gravitational redshift due to massive black holes etc). But even with the most "favourable" assumption the age of this cluster is reckoned to be only a few hundred million years at a time when galaxy formation would not have been remotely possible.

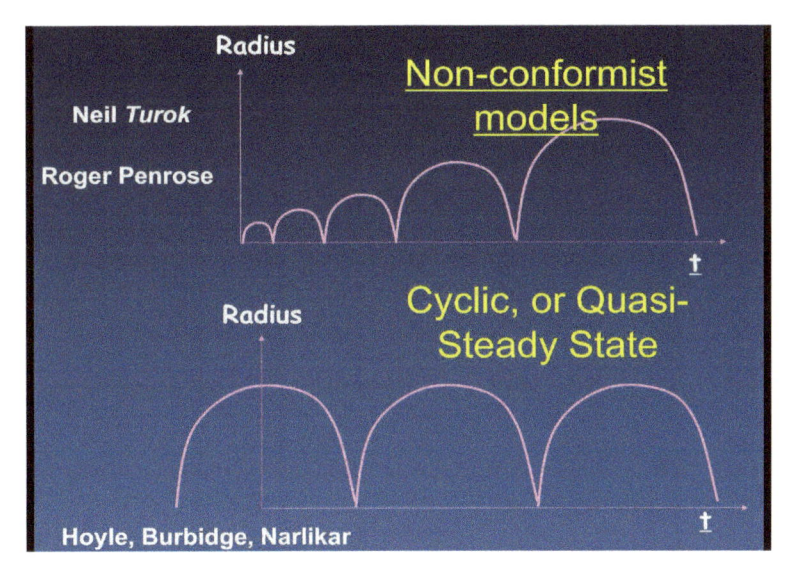

Fig. 1. A galaxy presumed to have formed according to standard cosmology in the first 250 million years post Big Bang – when atoms have scarcely formed! (courtesy NASA).

Fig 2. Schematic behaviour of quasi-steady state cosmological models (Penrose, 2010, Hoyle et al, 2000, Narlikar, 2002).

There are several alternative cosmological models/scenarios that attempt to explain the new results, although none of them have as wide an acceptance as the big bang theory, albeit in a modified and convoluted way. The Hoyle-Narlikar steady-state model of the universe posits the universe on the average always had and will always have the same average density. The theory reconciles the apparent evidence that the universe is expanding by suggesting that the universe generates matter at a rate proportionate to the universe's rate of expansion.

5. Conclusion

The long-held confidence in the uniqueness of the class of Big-Bang cosmologies appears to be seriously eroded by a wide range of new astronomical data, notable data from the deployment of the James Webb Space Telescope (JWST). Whether the resulting conflicts can be suitably resolved is left to be seen, but our assessment at present is that an alternative class of cosmologies with an open timescale, developed from the 1960's onward by Fred Hoyle and one to the present authors (JVN), needs to be seriously considered.

References

Arp, H.C., Burbidge, G., Hoyle, F., Narlikar, J.V., and Wickramasinghe, N.C.,1990. The extragalactic universe: an alternative view, *Nature*, **346**, 807

Bradley, L.D., *et al* 2023. High-redshift Galaxy Candidates at $z = 9–10$ as Revealed by JWST Observations of WHL0137-08, *Astrophys.J.*, **955** 13

Hoyle, F., Burbidge, G. and Narlikar, J.V., 2000. Alternative models of cosmology (Cambridge University Press)

Hubble, E., 1929. A relation between distance and radial velocity among extra-galactic nebulae". *Proceedings of the National Academy of Sciences* 15 (3): 168–173.

Narlikar, J.V. 2002. An Introduction to Cosmology (Cambridge University Press)

Penrose R. 2010. Cycles of Time: An Extraordinary New View of the Universe. (Bodley Head, London).

Steinhardt, P. J.; Turok, N. (2001). A Cyclic Model of the Universe, *Science*. **296** (5572): 1436–1439. arXiv:hep-th/0111030.

Epilogue

Burbidge, Burbidge, Fowler, Hoyle (B²FH) and Wickramasinghe
[and others]:

An Annotated Bibliography
with Notes on "Cosmology and the Origins of Life" [volume 30]
and Evolution,
including biological evolution, behavioral evolution, stellar evolution,
life as a cosmic phenomenon, cosmic life, panspermia, astrobiology,
cosmic biology, and cosmic evolution.

September 2023

Theodore Walker Jr., PhD
Southern Methodist University, Dallas, Texas 75275
Email: twalker@smu.edu

Contents -

Burbidge, E. Margaret
Burbidge, Geoffrey R.

Fowler, William A.

Hoyle, Fred

Hoyle, Fred, and **Nalin Chandra Wickramasinghe**
Hoyle, Fred, and Others

Wickramasinghe, Nalin Chandra
Wickramasinghe, Nalin Chandra, and Others

Wickramasinghe, **Dayal Tissa**, and Others
Wickramasinghe, **Janaki Tara**, and Others

Various Others on astronomy, cosmology, theology, origins of life, creation, and evolution
----

Burbidge, E. Margaret

Burbidge, E. Margaret, Geoffrey R. Burbidge, William A. Fowler, and Fred Hoyle. (1 October 1957). "Synthesis of the Elements in Stars" in *Review of Modern Physics*, volume 29, issue number 4, pages 547-650, doi:10.1103/RevModPhys.29.547, identified by author initials as ***B²FH***.

Concerning "synthesis of the elements" and "stellar evolution," see: "The Evolution of Stars" (Hoyle and Lyttleton 1939), "On the Accretion of Interstellar Matter by Stars" (Hoyle and Lyttleton 1940), "On the Accretion Theory of Stellar Evolution" (Hoyle and Lyttleton 1941), "On the Nature of Red Giant Stars" (Hoyle and Lyttleton 1942), "The Synthesis of the Elements from Hydrogen" (Hoyle 1946), "On the Condensation of the Planets" (Hoyle 1946), "The Chemical Composition of the Stars" (Hoyle 1946), "On the Formation of Heavy Elements in Stars" (Hoyle 1947), "A New Model for the Expanding Universe" (Hoyle 1948), "Stellar Evolution and the Expanding Universe" (Hoyle 1949), "On the Fragmentation of Gas Clouds into Galaxies and Stars" (Hoyle 1953), "On Nuclear Reactions Occurring in Very Hot Stars. I. The Synthesis of Elements from Carbon to Nickel" (Hoyle 1954), "On the Evolution of Type II Stars" (Hoyle and Schwarzschild 1955), "Origin of the Elements in Stars" (Hoyle, Fowler, Burbidge, and Burbidge 1956) and "Synthesis of the Elements in Stars" (Burbidge, Burbidge, Fowler, and Fred Hoyle [***B²FH***] 1957). Concerning "life as a cosmic phenomenon" and "cosmic evolution," see: "The Case for Life as a Cosmic Phenomenon" (Hoyle and Wickramasinghe 1986) and *Origins: Fourteen Billion Years of Cosmic Evolution* (Tyson and Goldsmith 2004) where ***B²FH*** is described as "a turning point in our knowledge of how the universe works" (page 165).

Burbidge, E. Margaret. (September 1994). "Watchers of the Skies" [a memoir expressing love of poetry with a title from Keats and "a lifetime in astronomy"] in *Annual Review of Astronomy and Astrophysics*, volume 32, pages 1-36, doi:10.1146/annurev.aa.32.090194.000245, https://adsabs.harvard.edu/pdf/1994ARA%26A..32....1B.

Burbidge, E. Margaret. (2005). "Modern Alchemy: Fred Hoyle and Element Building by Neutron Capture" is chapter 11 in *The Scientific Legacy of*

Fred Hoyle, edited by Douglas Gough. Cambridge: Cambridge University Press, 2011 paperback.

Burbidge, Geoffrey R.

Burbidge, Geoffrey R. (17 June 1961). "Galactic Explosions as Sources of Radio Emission" in *Nature*, volume 190, issue number 4781, pages 1053-1056, doi:10.1038/1901053a0. https://www.nature.com/articles/1901053a0.

Burbidge, Geoffrey R. (August 1966). "The Origin of Cosmic Rays" in *Scientific American*, volume 215, issue number 2, pages 32-38, doi:10.1038/scientificamerican0866-32, https://www.jstor.org/stable/24931020.

Burbidge, Geoffrey R. (1967). *Lectures on High Energy Astrophysics*. Bombay, India: Tata Institute of Fundamental Research.

Burbidge, Geoffrey R. (3 September 1971). "Was There Really a Big Bang?" in *Nature*, volume 233, issue number 5314, page 36-40, https://www.nature.com/articles/233036a0.

Burbidge, Geoffrey R. (12 November 1973). "Problem of the Redshifts" in *Nature Physical Science*, volume 246, issue number 150, pages 17-25, https://www.nature.com/articles/physci246017a0.

Burbidge, Geoffrey R. (February 1992). "Why Only One Big Bang?" in *Scientific American*, volume 266, page 120, https://www.scientificamerican.com/article/why-only-one-big-bang/.

Burbidge, Geoffrey R., FRS. (1 December 2003). "Sir Fred Hoyle. 24 June 1915 – 20 August 2001, Elected FRS 1957" in *Biographical Memoirs of Fellows of the Royal Society*, volume 49, pages 213-247, doi:10.1098/rsbm.2003.0013, https://royalsocietypublishing.org/doi/abs/10.1098/rsbm.2003.0013.

Burbidge, Geoffrey R. (2008). "B^{2}FH, the Cosmic Microwave Background and Cosmology" in *Publications of the Astronomical Society of Australia*, volume 25, issue number 1, pages 30-35, doi:10.1071/AS07029, https://doi.org/10.1071/AS07029.

Burbidge, Geoffrey R. (2008). "Hoyle's Role in B²FH" in *Science*, volume 319, issue number 5869, page 1484b, doi:10.1126/science.319.5869.1484b, https://www.jstor.org/stable/20053568.

Burbidge, Geoffrey R. (Submitted 14 November 2008). "A Realistic Cosmological Model Based on Observations and Some Theory Developed Over the Last 90 Years" [a paper presented at a meeting entitled "A Century of Cosmology" online 16 June 2008] in *Italian Physical Society/Società Italiana Di Fisica*, year 2007, issue number 12, pages 1437-1452, https://arxiv.org/pdf/0811.2402.pdf.

Burbidge, Geoffrey R., and **Fred Hoyle**. (December 1966). "The Problem of the Quasi-Stellar Objects" in *Scientific American*, volume 215, pages 40-52, doi:10.1038//Scientificamerican1266-40, https://www.jstor.org/stable/24931353.

Burbidge, Geoffrey R., and **E. Margaret Burbidge**. (1967). *Quasi-Stellar Objects*. San Francisco: W. H. Freeman Publisher.

Burbidge, Geoffrey R., and **Fred Hoyle**. (1 March 1969). "Condensed Objects in the Crab Nebula" in *Nature*, volume 221, issue number 5183, pages 847-848, doi:10.1038/221847a0, https://www.nature.com/articles/221847a0.

Burbidge, Geoffrey R., and **E. Margaret Burbidge**. (4 October 1969). "Quasi-stellar Objects—A Progress Report" in *Nature*, volume 224, issue number 5214, pages 21-24, doi:10.1038/224021a0, https://link.springer.com/article/10.1038/224021a0.

Fowler, William A.

Fowler, William A. (1954). "Experimental and Theoretical Results on Nuclear Reactions in Stars" in *Mémoires de la Societé Royale des Sciences de Liegè*, volume 13, pages 88-112.

Fowler, William A. (September 1956). "The Origin of the Elements" in *Scientific American*, volume 195, issue number 3, pages 82-91, doi:10.1038/scientificamerican0956-82, https://www.jstor.org/stable/24941744.

Fowler, William A. (July 1965). *Neutrino Astrophysics: Supermassive Stars, Quasars, and Extragalactic Radio Sources and Nuclear Energy Generation in Supermassive Stars*. W. K. Kellogg Foundation: Orange Aid Preprint Series in Nuclear Astrophysics.

Fowler, William A. (1967). *Nuclear Astrophysics*. Philadelphia, Pennsylvania: American Philosophical Society.

Fowler, William A. (November 1984). "The Quest for the Origin of the Elements" in *Science*, volume 226, issue number 4677, pages 922-935, doi:10.1126/science.226.4677.922, https://www.science.org/doi/abs/10.1126/science.226.4677.922.

Fowler, William A., and Jesse L. Greenstein. (15 April 1956). "Element-Building Reactions in Stars" in *Proceedings of the National Academy of Sciences*, volume 42, issue number 4, pages 173-180.

Fowler, William A., and **Fred Hoyle**. (November 1960 [21 May 1960]). "Nucleosynthesis in Supernovae" in *Astrophysical Journal*, volume 132, pages 565-590, https://adsabs.harvard.edu/full/record/seri/ApJ../0132/1960ApJ...132.. 565H.html, https://www.osti.gov/biblio/4099797.

Fowler, William A., and **Fred Hoyle**. (December 1964 [8 April 1964]). "Neutrino Processes and Pair Formation in Massive Stars and Supernovae" in *Astrophysical Journal Supplement Series*, volume 9, pages 201-319, doi:10.1086/190103, https://adsabs.harvard.edu/pdf/1964ApJS....9..201F.

Fowler, William A., and **Fred Hoyle**. (c1964). *Nucleosynthesis in Massive Stars and Supernovae*. Chicago: University of Chicago Press, 1965.

Hoyle, Fred

Hoyle, Fred. (1939). "Quantum Electrodynamics, part I and part II" in *Mathematical Proceedings of the Cambridge Philosophical Society*, volume 35, issue number 3, pages 419-462.

Hoyle, Fred. (2 December 1946 [6 April 1946]). "The Synthesis of the Elements from Hydrogen" in *Monthly Notices of the Royal Astronomical Society*, volume 106, issue number 5, pages 343-383.

Hoyle, Fred. (13 April 1946). "On the Condensation of the Planets" in *Monthly Notices of the Royal Astronomical Society*, volume 106, issue number 5, pages 406-422.

Hoyle, Fred. (29 December 1946). "The Chemical Composition of the Stars" in *Monthly Notices of the Royal Astronomical Society*, volume 106, issue number 4, pages 255-259.

Hoyle, Fred. (1947). "On the Formation of Heavy Elements in Stars" in *Proceedings of the Physical Society*, volume 59, issue number 6, pages 972-978.

Hoyle, Fred. (5 August 1948). "A New Model for the Expanding Universe" in *Monthly Notices of the Royal Astronomical Society*, volume 108, issue number 1748, pages 372-382.

Hoyle, Fred. (1949). *Some Recent Researches in Solar Physics*. Cambridge: Cambridge University Press.

Hoyle, Fred. (14 January 1949). "On the Cosmological Problem" in *Monthly Notices of the Royal Astronomical Society*, volume 109, issue number 3, pages 365-371.

Hoyle, Fred. (5 February 1949). "Stellar Evolution and the Expanding Universe" in *Nature*, volume 163, pages 196-198, doi:10.1038/163196a0, https://www.nature.com/articles/163196a0.

Hoyle, Fred. (1950). *The Nature of the Universe: A Series of Broadcast Lectures*. Oxford: B. Blackwell [New York: Harper, 1951].

Hoyle, Fred. (1952). *Nature of the Universe*. Oxford: Blackwell.

Hoyle, Fred. (1 June 1952). "Concepts of the Universe" in *New York Times Magazine*, pages 11-12, 50-51. [Kragh 1996: 463].

Hoyle, Fred. (1953). *A Decade of Decision*. London: W. Heinemann Publishing.

Hoyle, Fred. (15 August 1953). "Cosmic Origin of Radiation at Radio Frequencies" in *Nature*, volume 172, issue number 4372, pages 296-297, doi:10.1038/172296a0.
https://www.nature.com/articles/172296a0.

Hoyle, Fred. (November 1953). "On the Fragmentation of Gas Clouds into Galaxies and Stars" in *Astrophysical Journal*, volume 118, pages 513-528, doi:10.1086/145780,
https://adsabs.harvard.edu/pdf/1953ApJ...118..513H.

Hoyle, Fred. (13 March 1954). "Generation of Radio Noise by Cosmic Sources" in *Nature*, volume 173, issue number 4402, pages 483-484, doi:10.1038/173483a0, https://www.nature.com/articles/173483a0.

Hoyle, Fred. (September 1954 [Received 22 December 1953]). "On Nuclear Reactions Occurring in Very Hot Stars. I. The Synthesis of Elements from Carbon to Nickel" in *Astrophysical Journal Supplement*, volume 1, pages 121-146.

Hoyle, Fred. (1955). *Frontiers of Astronomy*. London: Heinemann Publishing; New York: Harper and Brothers Publishers [reprinted in 1956, 1961, and 1963; Spanish version in 1960, and 1970].

Hoyle, Fred. (7 May 1955). "The 'Horizon' of the Steady-State Universe" [reply to Thomas Gold 26 February 1955] in *Nature*, volume 175, issue number 4462, page 808, doi:10.1038/175808a0.
https://www.nature.com/articles/175808a0.

Hoyle, Fred. (September 1956). "The Steady-State Universe" in *Scientific American*, volume 195, pages 157-167, doi:10.1038/scientificamerican0956-157.

Hoyle, Fred. (1956). *Man and Materialism*. New York: Harper and Brothers Publishers.

Hoyle, Fred. (1957). *The Black Cloud*. London: Heinemann Publishing; New York: Harper and Brothers Publishers.

Hoyle, Fred. (1959). *Rockets in Ursa Major: A Novel*. London: Heinemann Publishing.

Hoyle, Fred. (1960). *The Nature of the Universe*. New York: Harper and Brothers Publishers.

Hoyle, Fred. (1960). "On the Origin of the Solar Nebula" in *Quarterly Journal of the Royal Astronomical Society*, volume 1, issue number 1, pages 28-55.

Hoyle, Fred. (1961). "Observational Tests in Cosmology" [44th Guthrie Lecture] in *Proceedings of the Physical Society*, volume 77, issue number 1, pages 1-16, doi:10.1088/0370-1328/77/1/302. https://iopscience.iop.org/article/10.1088/0370-1328/77/1/302/meta.

Hoyle, Fred. (1962). *Astronomy*. Garden City, New York: Doubleday Publishing Group.

Hoyle, Fred. (1963). "Formation of the Planets" in *Origin of the Solar System: Proceedings of a Conference Held at the Goddard Institute for Space Studies, New York, January 23-34, 1962*, edited by Robert Jastrow and A. G. W. Cameron. New York: Academic Press.

Hoyle, Fred. (1965). *Galaxies, Nuclei and Quasars*. New York: Harper and Row; London: Heinemann Publishing, 1966.

Hoyle, Fred. (9 October 1965). "Recent Developments in Cosmology" in *Nature*, volume 208, issue number 5006, pages 111-114, doi:10.1038/208111a0, https://link.springer.com/article/10.1038/208111a0.

Hoyle, Fred. (1966). *October the First is Too Late* [a science fiction novel]. London: Heinemann Publishing Company.

Hoyle, Fred. (1968). "Review of Recent Developments in Cosmology" [The Bakerian Lecture] in *Proceedings of the Royal Society of London, Series A. Mathematical and Physical Sciences*, volume 308, issue number 1492, pages 1-17, doi:10.1098/rspa.1968.0204, https://royalsocietypublishing.org/doi/abs/10.1098/rspa.1968.0204.

Hoyle, Fred. (1972). *From Stonehenge to Modern Cosmology*. San Francisco: W. H. Freeman Publishing.

Hoyle, Fred. (1973). *Nicolaus Copernicus: An Essay on His Life and Work*. London: Heinemann Publishing.

Hoyle, Fred. (1973). "The Origin of the Universe" in *Quarterly Journal of the Royal Astronomical Society*, volume 14, pages 278-287.

Hoyle, Fred. (1975). *Astronomy and Cosmology: A Modern Course*. San Francisco: W. H. Freeman Publisher.

Hoyle, Fred. (1975). *Astronomy Today*. London: Heinemann Publishing.

Hoyle, Fred. (1975). *Highlights in Astronomy*. San Francisco: W. H. Freeman Publisher.

Hoyle, Fred. (15 March 1975 [15 August 1974]). "On the Origin of the Microwave Background" in *The Astrophysical Journal*, volume 196, pages 661-670, doi:10.1086/153452, https://adsabs.harvard.edu/pdf/1975ApJ...196..661H.

Hoyle, Fred. (1977). *On Stonehenge*. London: Heinemann Publishing.

Hoyle, Fred. (1977). *Ten Faces of the Universe*. London: Heinemann Publishing.

Hoyle, Fred. (1978). *The Cosmogony of the Solar System*. Cardiff, Wales, United Kingdom: University College Cardiff Press.

Hoyle, Fred. (1980). *Steady-State Cosmology Revisited*. Cardiff, Wales, United Kingdom: University College Cardiff Press.

Hoyle, Fred. (15 April 1980). *The Relation of Biology to Astronomy*. Cardiff, Wales, United Kingdom: University College Cardiff Press.

Here the idea of life *as such* originating from some "warm little pond" (Darwin to Joseph Hooker 1871) or "primordial soup" (Oparin 1924; Haldane 1929) on planet Earth is replaced with the idea of life originating from a vastly larger pond, the Milky Way galaxy and billions of galactic ponds. Contrary to the astronomically improbable hypothesis that microbial life originated from our little pond; Fred Hoyle advanced the vastly more probable hypothesis that microbial life "did not begin on the Earth" (15 April 1980: 21), that "life is not confined to a particular galaxy," and that "Life can spread itself through the Universe" (15 April 1980: 23). Microbiology has a cosmic quality. Hoyle wrote: "*I suspect that the cosmic quality of microbiology will seem as obvious to future generations as the Sun being the centre of our solar system seems obvious to the present generation*" (15 April 1980: 24-25). [Italics added.] Also, concerning "cosmic biology," see "The Imperatives of Cosmic Biology" (2010) by Chandra Wickramasinghe and Carl H. Gibson, *Cosmic Biology: How Life Could Evolve on Other Worlds* (c2011) by Louis N. Irwin and Dirk Schulze-Makuch, "Growing Evidence for Cosmic Biology" (September 2014) by Chandra Wickramasinghe, Gensuke Tokoro, and Milton Wainwright, and *Vindication of Cosmic Biology: Tribute to Sir Fred Hoyle (2015-2001)* (2015) edited by Chandra Wickramasinghe.

Concerning **future generations** and cosmic biology, including astrobiology, in *The Relation of Biology to Astronomy* (Cardiff, Wales, United Kingdom: University College Cardiff Press, 15 April 1980) Fred Hoyle said, "I suspect that the cosmic quality of microbiology will seem as obvious to future generations as the Sun being the centre of our solar system seems obvious to the present generation" (24-25).
When is that future?
Predictions concerning when scientific evidence for extraterrestrial life will be widely received as obvious include the following years:
The year 2061 - In 2000, Arthur C. Clark foresaw probable compelling evidence in the year 2061. He said: "2061. The return of Halley's Comet; first landing by humans. The sensational discovery of both dormant and active life-forms vindicates Hoyle and Wickramasinghe's century old hypothesis that life is omnipresent throughout space." (Clarke 2000: 539)

By the year 2035 - In 2015, David Darling said, "Within the next 10 to 20 years there is every reason to hope that we will find the first evidence for life beyond Earth." (Darling's note to the second edition of - David Darling and Dick Schulze-Makuch 2016 [2000]).

By the year 2029 – In 2019, C. Wickramasinghe, K. Wickramasinghe, and G. Tokoro said: "We predict that ten years from now our cosmic origin will be deemed as obvious as the sun being the center of the solar system is considered obvious today" (2019: 1).

Perhaps sooner – Recent advances toward panpsychism in development biology (Levin 2019), mycology (Paul Stamets), and artificial intelligence (Gawdat 2021) might accelerate progress favoring panspermia, astrobiology, cosmic biology, and constructive postmodern natural scientific "astro-theology" (Derham 1715; Wickramasinghe and Walker 2015).

Hoyle, Fred. (May 1981). *The Universe: Past and Present Reflections.* Cardiff, Wales, United Kingdom: University College Cardiff Press.

Hoyle, Fred. (November 1981). "The Universe: Past and Present Reflections" in *Engineering and Science*, pages 8-12.

Hoyle, Fred. (12 November 1981). "Hoyle on Evolution" in *Nature*, volume 294, issue number 5837, pages 104-105.
Herein is Hoyle's famous "chance that a tornado sweeping through a junk-yard might assemble a Boeing 747" analogy for theory of evolution by chance assembly from random mutations.

Hoyle, Fred. (19 November 1981). "The Big Bang in Astronomy" in *New Scientist*, volume 92, issue number 1280, pages 521-527.

Hoyle, Fred. (1982). "The Universe: Past and Present Reflections" in *Annual Review of Astronomy and Astrophysics*, volume 20, pages 1-35, https://articles.adsabs.harvard.edu//full/1982ARA%26A..20....1H/0000001.000.html.

Hoyle, Fred. (December 1982). "From Virus to Cosmology."
Sir Fred Hoyle's IFS Lecture, audio-video online at https://www.buckingham.ac.uk/research/bcab/hrwarchive

Hoyle, Fred. (1984 [c1983]). *The Intelligent Universe: A New View of Creation and Evolution*. New York: Holt, Rinehart, and Winston Publishing.

Hoyle, Fred. (1985). *Comet Halley: A Novel in Two Parts*. New York: St. Martin's Press.

Hoyle, Fred. (1 January 1986). *The Small World of Fred Hoyle: An Autobiography*. London: M. Joseph Publisher.

Hoyle, Fred. (1989). "The Steady-State Theory Revived" in *Comments on Astrophysics*, volume 13, issue number 2, pages 81-86, https://adsabs.harvard.edu/pdf/1989ComAp..13...81H.

Hoyle, Fred. (4 May 1989). "Articles of Faith" [book review of *Science and Providence* (1989) by John C. Polkinghorne] in *Nature*, volume 339, issue number 6219, pages 23-24, doi:10.1038/339023a0, https://link.springer.com/content/pdf/10.1038/339023a0.pdf.

Hoyle, Fred. (5 October 1989). "What's in a Name?" in *Nature*, volume 341, issue number 6241, page 380-380.

Hoyle, Fred. (26 April 1990). "High Hopes for the Space Telescope" in *Nature*, volume 344, issue number 6269, pages 808-810, doi:10.1038/344808a0. https://www.nature.com/articles/344808a0.

Hoyle, Fred. (22 November 1990). "Birth of the Gods" in *Nature*, volume 348, issue number 6299, pages 353-354, doi:10.1038/348353a0, https://link.springer.com/content/pdf/10.1038/348353a0.pdf.

Hoyle, Fred. (15 April 1993). "Heavenly Works" [book review of *Nicholas Copernicus: Complete Works in Two Volumes* (1993) and *Johannes Kepler: New Astronomy* (1992)] in *Nature*, volume 362, issue number 6421, pages 657-658, doi:10.1038/362657a0, https://link.springer.com/content/pdf/10.1038/362657a0.pdf.

Hoyle, Fred. (1993). *The Origin of the Universe and the Origin of Religion*. Wakefield, Rhode Island: Moyer Bell Books.

Hoyle, Fred. (1994). *Home is Where the Wind Blows: Chapters from a Cosmologist's Life*. Mill Valley, California: University Science Books.

* Also, see "Preprints of Sir Fred Hoyle (1915-2001)," online at https://www.joh.cam.ac.uk/library/special_collections/personal_papers/hoylepreprints

Hoyle, Fred, and Nalin Chandra Wickramasinghe

Hoyle, Fred, and Nalin Chandra Wickramasinghe. (13 June 1962). "On Graphite Particles as Interstellar Grains" in *Monthly Notices of the Royal Astronomical Society*, volume 124, pages 417-433, https://academic.oup.com/mnras/article/124/5/417/2601371.

Hoyle, Fred, and Nalin Chandra Wickramasinghe. (8 July 1963). "On the Deficiency in the Ultraviolet Fluxes from Early Type Stars" in *Monthly Notices of the Royal Astronomical Society*, volume 126, issue number 4, pages 401-404, https://academic.oup.com/mnras/article/126/4/401/2602450.

Hoyle, Fred, and Nalin Chandra Wickramasinghe. (3 June 1967). "Impurities in Interstellar Grains" in *Nature*, volume 214, issue number 5092, pages 969-971, https://www.nature.com/articles/214969a0.

Hoyle, Fred, and Nalin Chandra Wickramasinghe. (1 February 1968). "Condensation of the Planets" in *Nature*, volume 217, issue number 5127, pages 415-418, doi:10.1038/217415a0, https://www.nature.com/articles/217415a0.

Hoyle, Fred, and Nalin Chandra Wickramasinghe. (22 June 1968). "Condensation of Dust in Galactic Explosions" in *Nature*, volume 218, issue number 5147, pages 1126-1127, http://physics.ruh.ac.lk/ab/pub/131.pdf.

Hoyle, Fred, and Nalin Chandra Wickramasinghe. (2 August 1969a). "Interstellar Graphite and Silicates" in *Nature*, volume 223, issue number 5205, page 445-446.
https://www.nature.com/articles/223445a0.

Hoyle, Fred, and Nalin Chandra Wickramasinghe. (2 August 1969b). "Interstellar Grains" in *Nature*, volume 223, issue number 5205, pages 459-462, doi:10.1038/223459a0,
https://www.nature.com/articles/223459a0.

Hoyle, Fred, and Nalin Chandra Wickramasinghe. (4 April 1970). "Dust in Supernova Explosions" in *Nature*, volume 226, issue number 5240, pages 62-63, https://www.nature.com/articles/226062a0.

Hoyle, Fred, and Nalin Chandra Wickramasinghe. (1 August 1970 [15 June 1970]). "Radio Waves from Grains in HII Regions" in *Nature*, volume 227, issue number 5257, pages 473-474, https://www.nature.com/articles/227473a0.

Hoyle, Fred, and Nalin Chandra Wickramasinghe. (4 November 1976). "Primitive Grain Clumps and Organic Compounds in Carbonaceous Chondrites" in *Nature*, volume 264, issue number 5581, pages 45-46, https://www.nature.com/articles/264045a0.

Hoyle, Fred, and Nalin Chandra Wickramasinghe. (1977). "Polysaccharides and the infrared spectra of galactic sources" in *Nature*, volume 268, page 610. [Darling and Schulze-Makuch 2016: 447]

Hoyle, Fred, and Nalin Chandra Wickramasinghe. (1 November 1977 [4 August 1977]). "Polysaccharides and the Infrared Spectrum of OH 26.5 + 0.6" in *Monthly Notices of the Royal Astronomical Society*, volume 181, *Short Communication*, pages 51P-55P, https://academic.oup.com/mnras/article/181/1/51P/1142216.

Hoyle, Fred, and Nalin Chandra Wickramasinghe. (24 November 1977). "Identification of the $\lambda 2$, 200Å Interstellar Absorption Feature" in *Nature*, volume 270, issue number 5635, pages 323-324, doi:10.1038/270323a0, https://www.nature.com/articles/270323a0.

Hoyle, Fred, and Nalin Chandra Wickramasinghe. (22 December 1977 [9 August 1977]) "Origin and Nature of Carbonaceous Material in the Galaxy" in *Nature*, volume 270, issue number 5639, pages 701-703, doi:10.1038/270701a0, https://www.nature.com/articles/270701a0.

Hoyle, Fred, Nalin Chandra Wickramasinghe, and A. H. Olavesen. (19 January 1978 [9 September 1977]) "Identification of Interstellar Polysaccharides and Related Hydrocarbons" in *Nature*, volume 271, issue number 5642, pages 229-231,
https://doi.org/10.1038/271229a0
https://www.nature.com/articles/271229a0.

Hoyle, Fred, and Nalin Chandra Wickramasinghe. (February 1978 [20 September 1977]). "Calculations of Infrared Fluxes from Galactic Sources for a Polysaccharide Grain Model" in *Astrophysics and Space Science*, volume 53, issue number 2, pages 489-505, https://link.springer.com/article/10.1007/BF00645036.

Hoyle, Fred, and Nalin Chandra Wickramasinghe. (February 1978 [19 December 1977]). "Comets, Ice Ages and Ecological Catastrophes" [Letter to the Editor] in *Astrophysics and Space Science*, volume 53, issue number 2, pages 523-526, https://link.springer.com/article/10.1007/BF00645040.

Hoyle, Fred, and Nalin Chandra Wickramasinghe. (1978). *Lifecloud: The Origin of Life in the Universe*. London: J. M. Dent & Sons.

Hoyle, Fred, and Chandra Wickramasinghe. (1979). *Diseases from Space*. London: J. M. Dent & Sons.

Hoyle, Fred, and Nalin Chandra Wickramasinghe. (30 March 1979). "On the Nature of Interstellar Grains" in *Astrophysics and Space Science*, volume 66, issue number 1, pages 77-90.
https://link.springer.com/chapter/10.1007/978-94-011-4297-7_26.

Hoyle, Fred, and Nalin Chandra Wickramasinghe. (May 1979). *On the Ubiquity of Bacteria: Searching the Planets and Beyond*. Astrophysics and Relativity Preprint Series 54, 51p

Hoyle, Fred, and Nalin Chandra Wickramasinghe. (May 1980 [6 November 1979]). "Organic Grains in Space" in *Astrophysics and Space Science*, volume 69, issue number 2, pages 511-513, https://link.springer.com/article/10.1007/BF00661935.

Hoyle, Fred, and Nalin Chandra Wickramasinghe. (1981 [29-31 October 1980]). "Comets—a Vehicle for Panspermia" (pages 227-239) in *Comets and the Origin of Life: Proceedings of the Fifth College Park Colloquium on Chemical Evolution, University of Maryland, College Park, Maryland, U.S.A., October 29th to 31st, 1980*, edited by Cyril Ponnamperuma. Boston: Kluwer Academic Publishers.

Wickramasinghe describes "Comets—a Vehicle for Panspermia" (Hoyle and Wickramasinghe: 1981 [29-31 October 1980]) as the first "explicit exposition of our cometary panspermia theory" in a reprint of this essay in *The Journal of Cosmology*, volume 16, September-October 2011.

Hoyle, Fred, and N. Chandra Wickramasinghe. (1981). *Evolution from Space: A Theory of Cosmic Creationism*. New York: Simon and Schuster.
The title of chapter 9 – "Convergence to God" is also the original more descriptive penultimate title of *the Big Bang and God: An Astro-Theology* ... (2015) by Theodore Walker Jr. and Chandra Wickramasinghe.

Hoyle, Fred, and N. Chandra Wickramasinghe. (1981). *Space Travellers: The Bringers of Life*. Cardiff, Wales, United Kingdom: University College, Cardiff Press.

Hoyle, Fred, and N. Chandra Wickramasinghe. (1982). "Comets" (pages 23-35) in *Proofs that Life is Cosmic*, Memoirs of the Institute of Fundamental Studies. Sri Lanka: Government Press.

Hoyle, Fred, and Nalin Chandra Wickramasinghe. (1982). *Why Neo-Darwinism Does Not Work*. Cardiff, Wales, United Kingdom: University College Cardiff Press.

Hoyle, Fred, and Nalin Chandra Wickramasinghe. (8 September 1983). "Organic Grains in Taurus Interstellar Clouds" in *Nature*, volume 305, issue number 5930, page 161, doi:10.1038/305161a0, https://www.nature.com/articles/305161a0.

Hoyle, Fred, and Nalin Chandra Wickramasinghe. (1 December 1983). "Bacterial Life in Space" in *Nature*, volume 306, issue number 5942, page 420, doi:10.1038/306420a0.
https://link.springer.com/content/pdf/10.1038/306420a0.pdf.

Hoyle, Fred, and Nalin Chandra Wickramasinghe. (1984). *From Grains to Bacteria*. Cardiff, Wales, United Kingdom: University College Cardiff Press.

Hoyle, Fred, Nalin Chandra Wickramasinghe, and Sirwan Al-Mufti. (1985 [29 October 1984]). "The Ultraviolet Absorbance of Presumably Interstellar Bacteria and Related Matters" in *Astrophysics and Space Science*, volume 111, issue number 1, pages 65-78.

Hoyle, Fred, Nalin Chandra Wickramasinghe, and Hans D. Plug. (July 1985 [11 February 1985]). "An Object within a Particle of Extra-terrestrial Origin Compared with an Object of Presumed Terrestrial Origin" [Letter to the Editor] in *Astrophysics and Space Science*, volume 113, issue number 1, pages 209-210, doi:10.1023/A:1002432332393, https://adsabs.harvard.edu/full/1985Ap%26SS.113..209H
Reprinted (pages 43-44) in *Astronomical Origins of Life: Steps towards Panspermia* (2000) edited by F. Hoyle and N. C. Wickramasinghe.

Hoyle, Fred, and Chandra Wickramasinghe. (1985). *Living Comets*. Cardiff, Wales, United Kingdom: University College Cardiff Press.

Hoyle, Fred, Nalin Chandra Wickramasinghe, and M. K. Wallis. (1985). "On the Nature of Dust Grains in the Comae of Comets Cernis and Bowell" in *Earth, Moon, and Planets*, volume 33, issue number 2, pages 179-187, https://link.springer.com/article/10.1007/BF00116794.

Hoyle, Fred, and N. Chandra Wickramasinghe. (1986). *Viruses from Space*. Cardiff, Wales, United Kingdom: University College Cardiff Press.

Hoyle, Fred, and Nalin Chandra Wickramasinghe. (7 August 1986). "The Case for Life as a Cosmic Phenomenon" in *Nature*, volume 322, issue number 6079, pages 509-511, doi:10.1038/322509a0, https://www.nature.com/articles/322509a0.

Hoyle, Fred, and Nalin Chandra Wickramasinghe. (9 July 1987). "Organic Dust in Comet Halley" in *Nature*, volume 328, issue number 6126, page 117, https://doi.org/10.1038/328117a0, https://www.nature.com/articles/328117a0#citeas.

Hoyle, Fred, and Nalin Chandra Wickramasinghe. (January 1988). "The Organic Nature of Cometary Grains" in *Earth, Moon, and Planets*, volume 40, issue number 1, pages 101-108, https://link.springer.com/article/10.1007/BF00057948.

Hoyle, Fred, and N. Chandra Wickramasinghe. (1988). *Cosmic Life-Force.* London: J. M. Dent & Sons; first American edition, New York: Paragon Books, 1990.
Includes "The Concept of a Creator" (pages 132-144).

Hoyle, Fred, and Nalin Chandra Wickramasinghe. (14 January 1988). "Cometary Organics" in *Nature,* volume 331, issue number 6152, pages 123-124, doi:10.1038/331123c0, https://www.nature.com/articles/331123c0.

Hoyle, Fred, and Nalin Chandra Wickramasinghe. (25 February 1988). "Cometary Organics" in *Nature,* volume 331, issue number 6158, page 666, https://www.nature.com/articles/331666c0.

Hoyle, Fred, and Nalin Chandra Wickramasinghe. (August 1988 [2 March 1988]). "Metallic Particles in Astronomy" in *Astrophysics and Space Science, volume* 147, pages 245–256, https://doi.org/10.1007/BF00645669, https://link.springer.com/article/10.1007/BF00645669#citeas.
Also, see "Metallic Particles in Astronomy" (October 1999) by F. Hoyle and N. C. Wickramasinghe.

Hoyle, Fred, and Chandra Wickramasinghe. (1990 [1988]). *Cosmic Life-Force.* New York: Paragon Books.
Includes "The Concept of a Creator" (pages 132-144).

Hoyle, Fred, and Nalin Chandra Wickramasinghe. (April 1990). "Influenza – Evidence Against Contagion" in *Journal of the Royal Society of Medicine,* volume 83, issue number 4, pages 258-261, https://journals.sagepub.com/doi/pdf/10.1177/014107689008300417.

Hoyle, Fred, and N. Chandra Wickramasinghe. (1991). *The Theory of Cosmic Grains.* Dordrecht, Netherlands: Kluwer Academic Publishers.

Hoyle, Fred, and N. Chandra Wickramasinghe. (1993). *Our Place in the Cosmos: The Unfinished Revolution.* London: J. M. Dent & Sons Publishing.

Hoyle, Fred, and Chandra Wickramasinghe. (1997). *Life on Mars? The Case for a Cosmic Heritage*, Foreword by series editor Paul R. Goddard. Redland, England: Clinical Press.

Hoyle, Fred, and Nalin Chandra Wickramasinghe. (October 1999). "Metallic Particles in Astronomy" in *Astrophysics and Space Science*, volume 268, pages 77–88, © 2000. Dordrecht, Netherlands: Kluwer Academic Publishers, https://doi.org/10.1023/A:1002492618280.
Also, see "Metallic Particles in Astronomy" (August 1988) by F. Hoyle and N. C. Wickramasinghe.

Hoyle, Fred, and Nalin Chandra Wickramasinghe. (October 1999). "The Universe and Life: Deductions from the Weak Anthropic Principle" in *Astrophysics and Space Science*, volume 268, issue numbers 1-3, pages 89-102, https://doi.org/10.1023/A:1002444702350.

Hoyle, F., and N. C. Wickramasinghe, editors. (2000 [1999]). *Astronomical Origins of Life: Steps towards Panspermia*. Dordrecht, Netherlands: Kluwer Academic Publishers.

Hoyle, Fred, and Others

Hoyle, Fred, and Raymond A. Lyttleton. (July 1939). "The Effect of Interstellar Matter on Climatic Variation" in *Mathematical Proceedings of the Cambridge Philosophical Society*, volume 35, issue number 3, pages 405-415, doi:10.1017/s0305004100021150. https://www.cambridge.org/core/journals/mathematical-proceedings-of-the-cambridge-philosophical-society/article/abs/effect-of-interstellar-matter-on-climatic-variation/0EA53316502FBA0B9D8FD21A62D7FF68.

Hoyle, Fred, and R. A. Lyttleton. (4 November 1939). "The Evolution of Stars" in *Mathematical Proceedings of the Cambridge Philosophical Society*, volume 35, issue number 4, page 592-609, https://doi.org/10.1017/S0305004100021368.

Hoyle, Fred, and R. A. Lyttleton. (June 1940). "On the Accretion of Interstellar Matter by Stars" in *Mathematical Proceedings of the Cambridge Philosophical Society*, volume 36, issue number 3, page 325-330, https://doi.org/10.1017/S0305004100017369.

Hoyle, Fred, and R. A. Lyttleton. (July 1940). "Note on Dr. Atkinson's Paper" in *Mathematical Proceedings of the Cambridge Philosophical Society*, volume 36, issue number 3, page 323-324, https://doi.org/10.1017/S0305004100017357.

Hoyle, Fred, and R. A. Lyttleton. (Received 8 March 1941). "On the Accretion Theory of Stellar Evolution" in *Monthly Notices of the Royal Astronomical Society*, volume 101, issue number 4, pages 227-236, https://academic.oup.com/mnras/article/101/4/227/2601180.

Hoyle, Fred, and R. A. Lyttleton. (Received 4 June 1942). "On the Nature of Red Giant Stars" in *Monthly Notices of the Royal Astronomical Society*, volume 102, issue number 5, pages 218-225, https://academic.oup.com/mnras/article/102/5/218/2600932.

Hoyle, Fred, D. Noel F. Dunbar, William A. Wenzel, and Ward Whaling. (January 1953). "A State in C^{12} Predicted from Astrophysical Evidence" in *Physical Review*, volume 92, issue number 4. page 1095-1095. [Kragh 2010: 34]

Hoyle, Fred, and M. Schwarzschild. (June 1955). "On the Evolution of Type II Stars" in *Astrophysical Journal Supplement*, volume 2, page 1, doi:10.1086/190015, https://ui.adsabs.harvard.edu/abs/1955ApJS....2....1H/abstract.

Hoyle, Fred, and Allan Sandage. (1956). "The Second-Order Term in the Redshift-Magnitude Relation" in *Proceedings of the Astronomical Society of the Pacific*, volume 68, issue number 403, pages 301-307, https://adsabs.harvard.edu/full/1956PASP...68..301H.

Hoyle, Fred, with William A. Fowler, G. R. Burbidge, and E. M. Burbidge. (5 October 1956). "Origin of the Elements in Stars" in *Science*, volume 124, issue number 3223, pages 611-614, doi:10.1126/science.124.3223.611, https://pubmed.ncbi.nlm.nih.gov/17832307/.

Hoyle, Fred, and Geoffrey Hoyle. (1959). *Rockers in Ursa Major* [a sci-fi novel]. London, Heinemann.

Hoyle, Fred, and J. V. Narlikar. (Received 13 June 1961). "On the Counting of Radio Sources in the Steady-State Cosmology" in *Monthly Notices of the Royal Astronomical Society*, volume 123, issue number 2, pages 133-149, https://academic.oup.com/mnras/article/125/1/13/2601372.

Hoyle, Fred, and John Elliot. (1962). *A for Andromeda: A Novel of Tomorrow*. London: Souvenir Press.
This 1962 sci-fi novel derives from the 1961 BBC black-and-white television series "A for Andromeda" written by John Elliot and Fred Hoyle. Elliot and Hoyle were attentive to dangers from artificial intelligence and biological engineering. "A for Andromeda" became a BBC color movie in 2006.

Hoyle, Fred, and J. V. Narlikar. (Received 5 July 1962). "On the Counting of Radio Sources in the Steady-State Cosmology, II" in *Monthly Notices of the Royal Astronomical Society*, volume 125, issue number 1, pages 13-20, https://adsabs.harvard.edu/full/1962MNRAS.125...13H.

Hoyle, Fred, and J. V. Narlikar. (February 1962). "The Steady-State Model and the Ages of Galaxies" in *Observatory*, volume 82, issue number 926, pages 13-14.
https://adsabs.harvard.edu/full/1962Obs....82...13H.

Hoyle, Fred, and William A. Fowler. (1 August 1962). "On the Nature of Strong Radio Sources" in *Monthly Notices of the Royal Astronomical Society*, volume 125, issue number 2, pages 169-176,
doi.org/10.1093/mnras/125.2.169,
https://adsabs.harvard.edu/full/1963MNRAS.125..169H.

Hoyle, Fred, and William A. Fowler. (9 February 1963). "Nature of Strong Radio Sources" in *Nature*, volume 197, issue number 4867, pages 533-535.

Hoyle, Fred, and Geoffrey Hoyle. (1963). *Fifth Planet* [science fiction]. New York: Harper and Row.

Hoyle, Fred, and Jayant Vishnu Narlikar. (7 January 1964). "Time Symmetric Electrodynamics and the Arrow of Time in Cosmology" in *Proceedings of the Royal Society of London*, volume 277, issue number 1368, pages 1-23, doi.org/10.1098/rspa.1964.0002,
https://royalsocietypublishing.org/doi/abs/10.1098/rspa.1964.0002.

Hoyle, Fred, and R. J. Tayler. (12 September 1964). "The Mystery of the Cosmic Helium Abundance" in *Nature*, volume 203, issue number 4950, pages 1108-1110,
doi:10.1038/2031108a0. https://www.nature.com/articles/2031108a0.

Hoyle, Fred, and J. V. Narlikar. (3 November 1964). "A New Theory of Gravitation" in *Proceedings of the Royal Society of London*, volume 282, issue number 1389, pages 191-207,
https://royalsocietypublishing.org/doi/pdf/10.1098/rspa.1964.0227.

Hoyle, Fred, and others. (1965). *University of Denver Centennial Symposium: The Responsible Individual and a Free Society in an Expanding Universe*. Denver, Colorado: Published for University of Denver by Big Mountain Press.

Hoyle, Fred, and J. V. Narlikar. (1966 [1 March 1965]). "A Radical Departure from the 'Steady-State' Concept in Cosmology" in *Proceedings of the Royal Society of London*, volume 290, issue number 1421, pages 162-176.

Hoyle, Fred, and William A. Fowler. (28 January 1967). "Gravitational Redshifts in Quasi-stellar Objects" in *Nature*, volume 213, issue number 5074, pages 373-374, doi:10.1038/213373a0. https://www.nature.com/articles/213373a0.

Hoyle, Fred, and Geoffrey R. Burbidge. (25 July 1970). "The Log S-log z Diagram for Radio Galaxies and its Relation to Cosmology" in *Nature*, volume 227, issue number 5256, pages 359-361, doi:10.1038/227359a0, https://www.nature.com/articles/227359a0.

Hoyle, Fred, and Geoffrey R. Burbidge. (1970). *Seven Steps to the Sun* [science fiction]. London: Heinemann.

Hoyle, Fred, and Geoffrey Hoyle. (1971). *The Molecule Men; and, The Monster of Loch Ness* [science fiction]. London, Heinemann.

Hoyle, Fred, and J. V. Narlikar. (January 1972 [27 July 1971]). "Cosmological Models in a Conformally Invariant Gravitational Theory—I: The Friedmann Models" in *Monthly Notices of the Royal Astronomical Society*, volume 155, issue number 3, pages 305-322, https://academic.oup.com/mnras/article/155/3/305/2603041.

Hoyle, Fred, and J. V. Narlikar. (January 1972 [27 July 1971]). "Cosmological Models in a Conformally Invariant Gravitational Theory—II: A New Model" in *Monthly Notices of the Royal Astronomical Society*, volume 155, issue number 3, pages 323-335, https://academic.oup.com/mnras/article/155/3/323/2603045.

Hoyle, Fred, and William A. Fowler. (9 February 1973 [Received 12 December 1972]). "On the Origin of Deuterium" in *Nature*, volume 241, issue number 5389, pages 384-386, doi:10.1038/241384a0. https://www.nature.com/articles/241384a0.

Hoyle, Fred, and Jayant V. Narlikar. (1974). *Action at a Distance in Physics and Cosmology*. San Francisco: W. H. Freeman Publisher.

Hoyle, Fred, and Jayant V. Narlikar, John Faulkner, editorial consultant. (1980). *The Physics-Astronomy Frontier*. San Francisco: W. H. Freeman.

Hoyle, Fred, Geoffrey R. Burbidge, and Jayant V. Narlikar. (June 1993). "A Quasi-Steady State Cosmological Model with Creation of Matter" in *Astrophysical Journal*, Part 1, volume 410, issue number 2, pages 437-457.

Hoyle, Fred, G. Burbidge, and J. V. Narlikar. (1994, [23 November 1993, original form 10 June 1993]). "Astrophysical Deductions from the Quasi-Steady-State Cosmology" in *Monthly Notices of the Royal Astronomical Society*, volume 267, pages 1007-1019, https://academic.oup.com/mnras/article/267/4/1007/1227351.

Hoyle, Fred, and Jayant V. Narlikar. (January 1995 [31 December 1994]). "Cosmology and Action-at-a-Distance Electrodynamics" in *Review of Modern Physics*, volume 67, issue number 1, pages 113-155, doi:10.1103/RevModPhys.67.113, https://journals.aps.org/rmp/abstract/10.1103/RevModPhys.67.113.

Hoyle, Fred, and Jayant V. Narlikar. (1996). *Lectures on Cosmology and Action at a Distance Electrodynamics*. River Edge, New Jersey: World Scientific Books.

Hoyle, Fred, G. Burbidge, and Jayant V. Narlikar. (1997 [21 June 1996]). "On the Hubble Constant and the Cosmological Constant" in *Monthly Notices of the Royal Astronomical Society*, volume 286, issue number 1, pages 173-182, doi:10.1093/mnras/286.1.173, https://academic.oup.com/mnras/article/286/1/173/1010536.

Hoyle, Fred, Geoffrey Burbidge, and Jayant V. Narlikar. (2000). *A Different Approach to Cosmology: From a Static Universe Through the Big Bang Towards Reality*. Cambridge: Cambridge University Press.

Wickramasinghe, Nalin Chandra

Wickramasinghe, Nalin Chandra. (Received 13 June 1962). "On Graphite Particles as Interstellar Grains" in *Monthly Notices of the Royal Astronomical Society*, volume 124, issue number 5, pages 417-433, https://academic.oup.com/mnras/article/124/5/417/2601371.

Wickramasinghe, N. C. (Received 8 February 1963 [originally 8 November 1962]). "On Graphite Particles as Interstellar Grains, II" in *Monthly Notices of the Royal Astronomical Society*, volume 126, issue number 1, pages 7, 99-114, https://academic.oup.com/mnras/article/126/1/99/2602427.

Wickramasinghe, N. C. (Received 4 January 1965). "On the Growth and Destruction of Ice Mantles on Interstellar Graphite Grains" in *Monthly Notices of the Royal Astronomical Society*, volume 131, issue number 2, pages 177-190, https://academic.oup.com/mnras/article/131/1/177/2604212.

Wickramasinghe, N. C. (1966 [Received 7 January 1965]). "On the Optics of Small Graphite Spheres, I" in *Monthly Notices of the Royal Astronomical Society*, volume 131, issue number 3, page 18, 263-269, https://academic.oup.com/mnras/article/131/2/263/2604198.

Wickramasinghe, N. Chandra. (1967). *Interstellar Grains*. London: Chapman & Hall Publishers.

Wickramasinghe, N. C. (15 November 1969). "Interstellar Polarization by Graphite-Silicate Grain Mixtures" in *Nature*, volume 224, issue number 5220, pages 656-658, doi:10.1038/224656a0, http://physics.ruh.ac.lk/ab/pub/94.pdf.

Wickramasinghe, N. C. (31 October 1970). "Between the Stars" in *Nature*, volume 228, issue number 5270, pages 483-484, doi:10.1038/228483c0, https://www.nature.com/articles/228483c0.

Wickramasinghe, N. C. (6 December 1974). "Formaldehyde Polymers in Interstellar Space" in *Nature*, volume 252, issue number 5483, pages 462-463, https://doi.org/10.1038/252462a0.

Wickramasinghe, Nalin Chandra. (April 1980). "The Origin of Life" is among the Historic Panspermia Lectures in the online Archive at the University of Buckingham, audio-video at https://www.buckingham.ac.uk/research/bcab/hrwarchive

Wickramasinghe, Nalin Chandra. (1982). *Is Life an Astronomical Phenomenon?* Cardiff, Wales, United Kingdom: University College Cardiff Press.

Wickramasinghe, N. C. (October 1999). "Formaldehyde Polymers in Interstellar Space" in *Astrophysics and Space Science, volume* 268, issue number 1, pages 111–114, https://doi.org/10.1023/A:1002448820097.

Wickramasinghe, Chandra. (2001). *Cosmic Dragons: Life and Death on Our Planet*. London: Souvenir Press.

Wickramasinghe, Chandra. (2004 [Received 29 September 2003]). "The Universe: A Cryogenic Habitat for Microbial Life" in *Cryobiology*, volume 48, pages 113-125.

Wickramasinghe, Chandra. (2005). "Alternative Cosmologies" is chapter 19 in *A Journey with Fred Hoyle: The Search for Cosmic Life*, edited by Kamala Wickramasinghe. London: World Scientific Books.

Wickramasinghe, Chandra. (2005). "From Dust to Life" is chapter 6 in *The Scientific Legacy of Fred Hoyle*, edited by Douglas Gough. Cambridge: Cambridge University Press, 2011 paperback.

Wickramasinghe, Chandra. (April 2010 [Online 29 January 2010]). "The Astrobiological Case for Our Cosmic Ancestry" in *International Journal of Astrobiology*, volume 9, issue number 2, pages 119-129, https://www.worldscientific.com/doi/abs/10.1142/9789814675260_0005.

Wickramasinghe, Chandra. (4 November 2010). "Microfossils in Comet Dust and Meteorites Support Panspermia" in *SPIE – the International Society for Optics and Photonics*, https://spie.org/news/3239-microfossils-in-comet-dust-and-meteorites-support-panspermia?SSO=1.

Wickramasinghe, Chandra. (2011). "The Compelling Case for Panspermia" (pages 211-224) in *Astronomy and Civilization in the New Enlightenment: Passions of the Skies* [*Analecta Husserliana: The Yearbook of Phenomenological Research, Volume CVII*], edited by Anna-Teresa Tymieniecka and Attila Grandpierre. Dordrecht, Neatherlands; London; New York: Springer Publishing.

Wickramasinghe, Nalin Chandra. (April 2011). "Extraterrestrial Life and Censorship." *Research Gate*, https://arxiv.org/abs/1104.1314, https://www.researchgate.net/publication/51021901_Extraterrestrial_Life_and_Censorship.

Wickramasinghe, Chandra. (September-October 2011). "From Astrochemistry to Astrobiology" in *The Journal of Cosmology*, volume 16, issue number 1, pages 6519-6527, https://thejournalofcosmology.com/Wick_1R%20-%20Copy.pdf.

Wickramasinghe, Nalin Chandra. (2 January 2014 [December 2011]). *A Destiny of Cosmic Life: Chapters in the Life of an Astrobiologist.* Kindle/Amazon Books.
Contains "A Galaxy Strewn with Microorganisms" (chapter 13) and "Cosmic Life, Microfossils and Evolution" (chapter 14).

Wickramasinghe, Nalin Chandra. (January 2013). "DNA Sequencing and Predictions of the Cosmic Theory of Life" in *Astrophysics and Space Science*, volume 343, issue number 1, pages 1-5, doi:10.1007/s10509-012-1227-y, https://link.springer.com/article/10.1007/s10509-012-1227-y.

Wickramasinghe, N. C. (February 2013). "Simulation of Earth-Based Theory with Lifeless Results" [Review of *First Life: Discovering the Connections between Stars, Cells, and How Life Began* (2012) by David Deamer] in *Bio Science*, volume 63, issue number 2, pages 141-143.

Wickramasinghe, Chandra. (2014). *The Search for Our Cosmic Ancestry.* Hackensack, New Jersey: World Scientific Publishing.

Wickramasinghe, N. Chandra. (September 2014). "Comet 67P/Churyumov-Gerasimenko and Cometary Biology" in *The Journal of Cosmology*, volume 24, issue number 3, 12032-12036, https://thejournalofcosmology.com/Rosetta_1.pdf.

Wickramasinghe, Chandra, editor. (2015). *Vindication of Cosmic Biology: Tribute to Sir Fred Hoyle (1915-2001)*. Hackensack, New Jersey: World Scientific.

Wickramasinghe, Chandra. (2015). *Where Did We Come from? Life of an Astrobiologist*, edited by Kamala Wickramasinghe. Hackensack, New Jersey: World Scientific.

Wickramasinghe, Chandra, with Introduction by Edward J. Steele. (14 July 2020). "COVID-19 Pandemic: A Challenge for Humanity" is Appendix 4 in *Diseases from Outer Space: Our Cosmic Destiny*. Hackensack, New Jersey: World Scientific.
Here is a revised and extended second edition of *Diseases from Space* (London: Dent, 1979) by Fred Hoyle and Chandra Wickramasinghe.

Wickramasinghe, Chandra. (May 2020). "Coronavirus from Outer Space - Interview with World-Renowned Astrobiologist Wickramasinghe" on YouTube at https://www.youtube.com/watch?v=o5LZ8YtCNz0. (Andre Waits, TC 8360, Summer 2020)

Wickramasinghe, Chandra. (May-June 2022). "Prof. Wickramasinghe – LIFE from SPACE – Our true origin"
on YouTube at https://www.youtube.com/watch?v=KveZopEdYa4.
Here Prof. Wickramasinghe says, "We are creatures of the cosmos … a cosmic living entity," and "all life" is "part of a single cosmic unity."

Wickramasinghe, Nalin Chandra, and Others

Wickramasinghe, N. C., and C. Guillaume (24 July 1965). "Interstellar Extinction by Graphite Grains" in *Nature*, volume 207, issue number 4995, pages 366-368, doi:10.1038/207366a0, http://www.physics.ruh.ac.lk/ab/pub/111.pdf.

Wickramasinghe, N. C., F. Hoyle, and K. Nandy. (1977). "Organic Molecules in Interstellar Dust: A Possible Spectral Signature at λ2200 Å?" [Letter to the Editor] in *Astrophysics and Space Science*, [10-1999] volume 268, issue number 1-3, pages 295-299 [volume 47, L9-L11].

Wickramasinghe, N. C., F. Hoyle, and K. Nandy. "Organic Molecules in Interstellar Dust: A Possible Spectral Signature at λ2200 Å?" in *Astronomical Origins of Life* (Dordrecht: Springer, 2000), pages 295-299, https://link.springer.com/chapter/10.1007/978-94-011-4297-7_30.

Wickramasinghe, N. C., F. Hoyle, J. Brooks, and G. Shaw. (20 October 1977 [19 August 1977]). "Prebiotic Polymers and Infrared Spectra of Galactic Sources" [Letter to Nature] in *Nature*, volume 269, issue number 5630, pages 674-676, doi:10.1038/269674a0, https://www.nature.com/articles/269674a0.

Wickramasinghe, N. C., and J. V. Narlikar. (7 October 1967). "Microwave Background in a Steady-State Universe" in *Nature*, volume 216, issue number 5110, pages 43-44, doi:10.1038/216043a0, https://www.nature.com/articles/217339a0.

Wickramasinghe, N. C., and K. S. Krishna Swamy. (1968 [Received 13 November 1967]). "On the Temperature of Interstellar Grains" in *Monthly Notices of the Royal Astronomical Society*, volume 139, issue number 3, pages 283-287.

Wickramasinghe, N. C., J. G. Ireland, K. Nandy, H. Seddon, and R. D. Wolstencroft. (3 February 1968). "Origin of the Diffuse Interstellar Bands" in *Nature*, volume 217, issue number 5127, pages 412-415, doi:10.1038/217412b0, https://www.nature.com/articles/217412b0.

Wickramasinghe, N. C., and A. H. Olavesen. (26 October 1978). "Cosmochemistry and Evolution" in *Nature*, volume 275, issue number 5682, page 694, doi:10.1038/275694a0, https://www.nature.com/articles/275694a0.

Wickramasinghe, N. C., A. N. Wickramasinghe, and Fred Hoyle. (October 1992). "The Case Against Graphite Particles in Interstellar Space" [Letter to the Editor] in *Astrophysics and Space Science*, volume 196, issue number 1, pages 167-169.
Also, in *Astronomical Origins of Life: Steps towards Panspermia* (Dordrecht: Springer, 2000), edited by F. Hoyle and N. C. Wickramasinghe, pages 289-292.

Wickramasinghe, N. C., and Daisaku Ikeda. (1998). *Space and Eternal Life: A Dialogue between Chandra Wickramasinghe and Daisaku Ikeda*, Foreword by Sir Fred Hoyle. Chicago: Journeyman Press.

Wickramasinghe, Chandra, Jayant Narlikar, and Geoffrey Burbidge, editors. (2003). *Fred Hoyle's Universe*. Dordrecht, Netherlands: Kluwer Academic Publishers [from Proceedings of a Conference Celebrating Fred Hoyle's Extraordinary Contributions to Science 25-26 June 2002 Cardiff University, United Kingdom].

Wickramasinghe, N. C., J. V. Narlikar, J. T. Wickramasinghe and M. Wainwright. (2003). "The Expanding Horizons of Cosmic Life" in *Proceedings of SPIE – International Society for Optics and Photonics*, volume 4859, pages 154-163, https://doi.org/10.1117/12.459301.

Wickramasinghe, N. Chandra, and Carl H. Gibson. (2 March 2010 [Submitted 27 February 2010]). "The Imperatives of Cosmic Biology" in *Inspire: High Energy Physics Information System*, 17 pages, https://arxiv.org/abs/1003.0091, https://www.researchgate.net/publication/45903478_The_Imperatives _of_Cosmic_Biology.

Wickramasinghe, N. Chandra, and Janaki T. Wickramasinghe. (5 June 2012). "A Note on Venus Transit and Microbial Injection to Earth" in *The Journal of Cosmology*, volume 18, issue number 20, pages 8506-8510, https://thejournalofcosmology.com/VENUS2012_R.pdf.

Wickramasinghe, N. C., J. Wallis, D. H. Wallis, and A. Samaranayake. (10 January 2013). "Fossil Diatoms in a New Carbonaceous Meteorite" in *The Journal of Cosmology*, volume 21, issue number 37, pages 9560-9571, https://thejournalofcosmology.com/PolonnaruwaRRRR.pdf.

Wickramasinghe, N. C., J. Wallis, D. H. Wallis, M. K. Wallis, S. Al-Mufti, J. T. Wickramasinghe, A. Samaranayake and K. Wickramarathne. (13 January 2013). "On the Cometary Origin of the Polonnaruwa Meteorite" in *The Journal of Cosmology*, volume 21, issue number 38, pages 9572-9578, https://thejournalofcosmology.com/Polonn2.pdf.

Wickramasinghe, N. C., J. Wallis, D. H. Wallis, M. K. Wallis, N. Miyake, S. G. Coulson, Carol H. Gibson, J. T. Wickramasinghe, A. Samaranayake, K. Wickramarathne, and Richard B. Hoover. (6 March 2013). "Incidence of Low-Density Meteoroids of the Polonnaruwa-Type" in *The Journal of Cosmology*, volume 22, issue number 1, pages 1-8, https://thejournalofcosmology.com/Paper22(1a).pdf.

Wickramasinghe, Chandra, and Gensuke Tokoro. (January 2014). "Life as a Cosmic Phenomenon: 2. The Panspermia Trajectory of *Homo sapiens*" in *Astrobiology and Outreach*, volume 2, issue number 2, 115, https://www.walshmedicalmedia.com/open-access/life-as-a-cosmic-phenomenon-the-panspermia-trajectory-of-homo-sapiens-2332-2519-2-115.pdf.

Wickramasinghe, N. Chandra, Gensuke Tokoro, and Milton Wainwright. (September 2014). "Growing Evidence for Cosmic Biology" in *The Journal of Cosmology*, 2014, volume 24, issue number 8, pages 12097-12101, https://thejournalofcosmology.com/Cosmic%20Biology%20rev.pdf. Abstract - New data from astronomy and biology continues to favour the Hoyle-Wickramasinghe theory of cometary panspermia. Alternative explanations on the basis of Earth-centred biology, with Neo-Darwinian evolution occurring within a closed system, appear to be far-fetched and fundamentally flawed. Keywords: Astrobiology, panspermia, interstellar matter, viruses.

Wickramasinghe, N. Chandra, Gensuke Tokoro, and Milton Wainwright. (September 2014). "The Transition from Earth-Centred Biology to Cosmic Life" [a paper presented at United Nations/Austria Symposium on "Space Science and the United Nations" Graz, Austria, 22 to 24 September 2014] in *The Journal of Cosmology*, 2014, volume 24, issue number 7, pages 12080-12096, https://thejournalofcosmology.com/CHANDRA%20PAPER-RR4.pdf.

Wickramasinghe, Chandra, and Gensuke Tokoro. (2015 [2014]). "Life as a Cosmic Phenomenon: 1. The Socio-economic Control of a Scientific Paradigm" and "Life as a Cosmic Phenomenon: 2. The Panspermic Trajectory of Homo Sapiens" (pages 3-33) in *Vindication of Cosmic Biology: Tribute to Sir Fred Hoyle (2015-2001)* (2015) edited by Chandra Wickramasinghe. Hackensack, New Jersey: World Scientific.

Wickramasinghe, Chandra, and Theodore Walker Jr., with editing by Alexander Vishio. (2015). *The Big Bang and God: An Astro-Theology wherein an astronomer and a theologian offer a study of interdisciplinary convergences with natural theology both in the scientific researches of Sir Fred Hoyle and in the philosophical researches of Charles Hartshorne and Alfred North Whitehead, thereby illustrating a constructive postmodern trend.* New York: Palgrave Macmillan.

Wickramasinghe, Chandra, and Robert Bauval. (2017). *Cosmic Womb: The Seeding of Planet Earth*. Rochester, Vermont: Bear & Company Books.

Wickramasinghe, Chandra, Kamala Wickramasinghe, and Gensuke Tokoro. (2019). *Our Cosmic Ancestry in the Stars: The Panspermia Revolution and the Origins of Humanity*. Rochester, Vermont: Bear & Company Books.

Wainwright, Milton, and N. Chandra Wickramasinghe, with Foreword by Gensuke Tokoro. (2023). *Life Comes from Space: The Decisive Evidence*. Hackensack, New Jersey: World Scientific.

Wickramasinghe, N. Chandra, and others. (Begun March 2023).
"Cosmology and the Origins of Life"
in *The Journal of Cosmology*, volume 30,
online at
https://thejournalofcosmology.com/indexVol30CONTENTS.htm:

1. N. Chandra Wickramasinghe, Jayant V. Narlikar and Gensuke Tokoro, Cosmology and the Origins of Life, New evidence related to the origins of life in the cosmos combined with continuing progress in probing conditions of the early universe using the James Web Telescope suggest that long-held orthodox positions may be flawed. Only by objective evaluating the new facts and recognising the cultural forces at work can further progress be made towards resolving perhaps the most important and fundamental questions in science. pp 30001 - 30013.

2. N. Chandra Wickramasinghe Life Beyond the Limits of Our Planetary System, Evidence for the widespread distribution of biologically relevant molecules widely throughout the Galaxy and beyond has been in existence for many decades. The recent discovery of a nucleobase uracil adds to an already impressive body of evidence that supports a cosmic origin of the complex building blocks of life. pp 30020 - 30024.

3. N. Chandra Wickramasinghe and Gensuke Tokoro, Quest for Life on Jupiter and Its Moons, The final confirmation of the existence of multicellular life in aqueous habitats on the moons of Jupiter, will be a game changer for the societal approval and acceptance of panspermia which has been long overdue. pp 30030 - 30034.

4. N. Chandra Wickramasinghe, Gensuke Tokoro, Robert Temple and Rudy Schild Reluctance to Admit We Are Not Alone as an Intelligent Lifeform in the Cosmos, With an ever-increasing body of evidence from diverse scientific disciplines all pointing to the existence of alien life and alien intelligence on a cosmic scale, there has developed a growing tendency to maintain that we might still be alone as intelligent beings in the universe. This a stubborn resistance to admit facts may well signal the end of our civilization. pp 30040 - 30053.

5. N. Chandra Wickramasinghe, Rudy Schild and J.H. (Cass) Forrington Second Copernican Revolution, The recent discovery by the James Webb Space Telescope of organic molecules possibly related to life in

a galaxy at redshift z=12.4 may well signal a concluding phase of the second Copernican revolution, thus removing the Earth from the centre and focus of biology and charting a new course in our understanding of the universe, and concluding a process that began 4 decades ago. pp 30060 - 30071.

6. N. Chandra Wickramasinghe, Rudy Schild, Gensuke Tokoro, Robert Temple and J.H. (Cass) Forrington Search For Aliens, and UFO's, The widespread existence of primitive life in the form of bacteria and viruses in the universe combined with the large numbers of habitable planets that are being discovered, leads to the serious possibility that intelligent life could be widespread throughout the cosmos. Discovering such alien intelligence in our vicinity continues to pose a challenge. pp 30080 - 30089.

Wickramasinghe, **Dayal Tissa**, and Others

Wickramasinghe, Dayal Tissa, and David A. Allen. (9 October 1980). "The 3.4-μm Interstellar Absorption Feature" in *Nature*, volume 287, issue number 5782, pages 518-519, doi:10.1038/287518a0, https://www.sciencedirect.com/science/article/abs/pii/S0065266020300067.

Wickramasinghe, Dayal Tissa, Fred Hoyle, Nalin Chandra Wickramasinghe, and Sirwan Al-Mufti. (Received 24 June 1986). "A Model of the 2-4 Micron Spectrum of Comet Halley" [Letter to the Editor] in *Earth, Moon, and Planets*, volume 36, pages 295-299.

Wickramasinghe, Dayal Tissa, and David A. Allen. (4 September 1986). "Discovery of Organic Grains in Comet Halley" in *Nature*, volume 323, issue number 6083, pages 44-46, doi:10.1038/323044a0, https://www.nature.com/articles/323044a0.

Wickramasinghe, Dayal Tissa, and David A. Allen. (21 October 1987). "Discovery of Organic Grains in Comet Wilson" in *Nature*, volume 329, issue number 6140, pages 615-616, doi:10.1038/329615a0, https://www.nature.com/articles/329615a0.

Wickramasinghe, **Janaki Tara**, and Others

Wickramasinghe, Janaki Tara. (25 May 2007). *The Role of Comets in Panspermia*: Cardiff University PhD Thesis. Ann Arbor, Michigan: ProQuest LLC Publishing, 2013 [MUI Dissertation Publishing, UMI Issue number: U584951].

Wickramasinghe, J. T., and N. C. Wickramasinghe. (December 2006). "A Cosmic Prevalence of Nanobacteria?" in *Astrophysics and Space Science*, vol. 305, issue number 4, pages 411-413 [Also in *Vindication of Cosmic Biology* (2015) pages 193-198], doi:10.1007/s10509-006-9181-1, https://link.springer.com/article/10.1007/s10509-006-9181-1.

Wickramasinghe, Janaki, Chandra Wickramasinghe, and William Napier. (2010 [July 2009]). *Comets and the Origin of Life*. Hackensack, New Jersey: World Scientific Books.

Various Others on astronomy, cosmology, theology, origins of life, creation, and evolution

Baker-Fletcher, Karen. (1998). *Sisters of Dust, Sisters of Spirit: Womanist Wordings on God and Creation*. Minneapolis: Fortress Press.

Clark, Arthur C. (2000 [1999]). "The Twenty-First Century: A (Very) Brief History" in *Greetings, Carbon-based Bipeds!: Collected Essays 1934-1998*, edited by Ian T. Macauley. New York: St. Martin's.

Coulson, S. G., and N. C. Wickramasinghe. (August 2003 [online 12 August 2003]). "Frictional and Radiation Heating of Micro-Sized Meteoroids in the Earth's Upper Atmosphere" in *Monthly Notices of the Royal Astronomical Society*, volume 343, issue number 4, pages 1123-1130, https://academic.oup.com/mnras/article/343/4/1123/1065355.

Darling, David, and Dick Schulze-Makuch. (2016 [2000]). *The Extraterrestrial Encyclopedia: An Alphabetic Guide to Life in the Universe*. Sarasota, Florida: First Edition Design Publishing.

Davis, Andrew M. (2023). *Metaphysics of Exo-Life: Toward a Constructive Whiteheadian Cosmotheology*. Grasmere, ID: SacraSage Press.

Deamer, David. (2012 [c2011]). *First Life: Discovering the Connections between Stars, Cells, and How Life Began*. Berkeley: University of California Press.

Derham, William. (1715). *Astro-Theology: or, A Demonstration of the Being and Attributes of God, from a Survey of the Heavens*. London: Printed for William Innys.
See *excerpts* from Derham's *Astro-Theology* ... (1715) in the Astro-Theology volume of *The Journal of Cosmology*, volume 20, pages 8746-9749, online at https://thejournalofcosmology.com/Derham%20excerpts%201.pdf. Also, see *The Big Bang and God*: *An Astro-Theology* ... (2015) by Theodore Walker Jr. and Chandra Wickramasinghe.

Devenish, Philip E. (4 October 1981). "Mind, Brain, and Dualism" in *The Journal of Religion*, volume 61, number 4, pages 422-427.

Devenish, Philip E., and George L. Goodwin, editors. (1989). *Witness and Existence: Essays in Honor of Schubert M. Ogden*. Chicago: University of Chicago Press.

Dover, Cedric. (June 1954). "The Significance of the Cell Surface (The Work of E. E. Just)" in *Journal of the Zoological Society of India*, volume 6, issue number 1, pages 3-42.

In India, said Cedric Dover [Cedric Cyril Dover], "we have preserved in biology a view of life as process, as interrelatedness, as a totality of subtle harmonies rather than a Darwinian war ..." and "Asian biologists" (including Calcutta zoologists such as Nelson Annandale and Sunder Lal Hora) emphasize "living things in their natural milieu." Accordingly, concerning Howard University biologist Ernest Everett Just, Dover said: "He belongs to their company ... his philosophy, like theirs, was a unitary one ..." (June 1954: 3-4).

Also, concerning Cedric Dover, biology, race, color, and class in the USA and India, see *The Prism of Race: W.E.B. Du Bois, Langston Hughes, Paul Robeson, and the Colored World of Cedric Dover* (Palgrave Macmillan, 2014) by Nico Slate.

Also, by Nico Slate:
Colored Cosmopolitanism: The Shared Struggle for Freedom in the United States and India (Harvard University Press, 2012), *Gandhi's Search for the Perfect Diet: Eating with the World* (University of Washington Press, 2019), *Lord Cornwallis Is Dead: The Struggle for Democracy in the United States and India* (Harvard University Press, 2019), and *Brothers: A Memoir of Love, Loss, and Race* (Temple University Press, 2023).

Gawdat, Mo. (2021). *Scary Smart: The Future of Artificial Intelligence and How You Can Save Our World*. Bluebird | Pan Macmillan.

Gibson, Carl H., Rudolph E. Schild, and N. Chandra Wickramasinghe. (2010 [Submitted 4 April 2010, accepted 9 August 2010]). "The Origin of Life from Primordial Planets" in *International Journal of Astrobiology*, volume 10, issue number 2, pages 83-98, doi:10.1017/S1473550410000352, https://arxiv.org/abs/1004.0504.

Gingerich, Owen. (2014). Chapter 3 "Was Hoyle Right?" in *God's Planet*. Cambridge, Massachusetts: Harvard University Press.

Gough, Douglas, editor. (2005). *The Scientific Legacy of Fred Hoyle.* Cambridge: Cambridge University Press, 2011 paperback.

Graves, Joseph L. (2022). *A Voice in the Wilderness: A Pioneering Biologist Explains How Evolution Can Help Us Solve Our Biggest Problems.* New York: Basic Books.

Gregory, Jane. (January 2003). "The Popularization and Excommunication of Fred Hoyle's 'Life-from-Space' Theory" in *Public Understanding of Science*, volume 12, issue number 1, pages 25-46.

Gregory, Jane. (2005). *Fred Hoyle's Universe.* Oxford: Oxford University Press.

Grevesse-Guillaume, C., and N. C. Wickramasinghe. (1966 [Received 21 July 1965]). "On the Optics of Small Graphite Spheres, III" in *Monthly Notices of the Royal Astronomical Society*, volume 132, issue number 4, pages 471-473, https://adsabs.harvard.edu/pdf/1966MNRAS.132..471G.

Haldane, J. B. S. [John Burdon Sanderson Haldane] (1929). *The Origin of Life.* London: Chatto and Windus Publishing.

Hannay, J. B. (15 February 1883). "Natural Selection and Natural Theology" [Letter to the Editor] in *Nature*, volume 27, issue number 694, page 364-364, doi:10.1038/027364a0, https://www.nature.com/articles/027364a0.

Hartshorne, Charles. (c1937). *Beyond Humanism: Essays in the Philosophy of Nature.* Chicago; New York: Willett, Clark & Company [Reprinted with important new preface, Gloucester, Massachusetts: Peter Smith Books, 1975].
  Concerning relations between astronomy and biology: In 1936-37 in *Beyond Humanism: Essays in the Philosophy of Nature* (1975 [c1937]) Charles Hartshorne was saying "astronomy is not as yet of much help in determining the prevalence in space-time of conditions favoring animal organism" (58). Since then, especially since B^2FH (1957), astronomers have learned to be of much help to

biology; and in so doing, they created the new convergent discipline of *astrobiology*.]

Hartshorne, Charles. ([1941a] 1964). "The Theological Analogies and the Cosmic Organism" (chapter V, pages 174-211) in *Man's Vision of God and the Logic of Theism*. Hamden, Connecticut: Archon Books, 1964 [New York: Willet, Clark & Company, 1941].

Hartshorne, Charles. (1991). "An Open Letter to Carl Sagan" in *The Journal of Speculative Philosophy*, volume 5, pages 227-232.

Hartshorne, Charles, and William L. Reese. (1953). *Philosophers Speak of God*. Chicago: University of Chicago Press. Reprints: Chicago: Midway Reprints, 1976; Amherst, New York: Humanity Books, 2000. [Here Alfred North Whitehead's view of God is classified as "panentheism." The word "panentheism" comes from pan-en-theos-ism, literally meaning <all included-in God> -ism (which is distinguished from classical theism, and from classical pantheism).]

Haselgrove, C. B., and F. Hoyle. (10 May 1956). "A Preliminary Determination of the Age of Type II Stars" in *Monthly Notices of the Royal Astronomical Society*, volume 116, issue number 5, pages 527-532.

Hauerwas, Stanley. (2001). *With the Grain of the Universe: The Church's Witness and Natural Theology* – Being the Gifford Lectures Delivered at the University of St. Andrews in 2001. Grand Rapids, Michigan: Brazos Press.

Hoover, Richard B. (July-August 2011). "Fossils of Cyanobacteria in CI1 Carbonaceous Meteorites: Implications to Life on Comets, Europa, and Enceladus" in *The Journal of Cosmology*, volume 15, issue number II-3, pages 6249-6287, https://thejournalofcosmology.com/Contents15_files/Hoover_JOC_MS.pdf.

Hoover, Richard B., Fred Hoyle, Nalin Chandra Wickramasinghe, Miriam J. Hoover, and Sirwan Al-Mufti. (April 1986). "Diatoms on Earth, Comets, Europa, and in Interstellar Space" in *Earth, Moon and Planets*, volume 35, issue number 1, pages 19-45.

https://link.springer.com/chapter/10.1007/978-94-011-4297-7_23.

Hoover, Richard B., A. Yu Rozanov, and Roland Paepe, editors. (2005). *Perspectives in Astrobiology*, Amsterdam: IOS Press.

Irwin, Louis N., and Dirk Schulze-Makuch. (c2011). *Cosmic Biology: How Life Could Evolve on Other Worlds.* New York; London: Springer; Chichester, UK; Published in association with Praxis Pub.

Jenkins, Lillie R. (3 April 2021). "Black Apollo of Science: The Life of Ernest Everett Just – Summarizing Timeline, Sumitography and Concept Poster" [Alternative title: "E. E. Just: Administrative and Fund-seeking Pioneer"], online via *SMU Scholar* at https://scholar.smu.edu/theology_research/27/.

Just, Ernest Everett. (January 1939). *The Biology of the Cell Surface.* Philadelphia: P Blakiston's Son.

Just, Ernest Everett. (June 1939). *Basic Methods for Experiments on Eggs of Marine Animals.* Philadelphia: P. Blakiston's Son.

Just, Ernest Everett, and Hedwig Schnetzler Just. (October 1941 unpublished [2020]). *The Origin of Man's Ethical Behavior.*
This 251-page archival edition was transcribed and edited during 2018-2020 by Theodore Walker Jr. and Lillie R. Jenkins, with additional co-editing by Walker, Jenkins, and W. Malcolm Byrnes, in consultation with Stuart Newman, Kenneth R. Manning, Charles H. Long, and Joellen ElBashir; independently published as *The ORIGIN OF MAN'S ETHICAL BEHAVIOR (1941) by ERNEST EVERETT JUST & HEDWIG SCHNETZLER JUST.*

Joseph, Gabriel R., Chandra Wickramasinghe, Richard Hoover, Gilbert Levin, Ben Goertzel, Allan Combs, Robert J.D. McLean, Malcolm A. C. McLean, Milton Wainwright, Pabulo Herique Rampelotto, and others. (18 November 2017). *Aliens, Extraterrestrials, Space Fungi, Moon Microbes, Martian Mushrooms, Diseases from Space, Sagan's Aliens in the Thermosphere, Evolution.* Cambridge, MA.: Cosmology Science Publishers.

Joseph, Rhawn, and N. Chandra Wickramasinghe. (September-October 2011). "Genetics Indicates Extraterrestrial Origins for Life: The First Gene – Did Life Begin Following the Big Bang?" in *The Journal of Cosmology*, volume 16, issue number 21, https://thejournalofcosmology.com/27_JosephWickGeneticOriginsLife.pdf.

Karim, L. M., Fred Hoyle, and N. C. Wickramasinghe (July 1983). "Interstellar Proteins and the Discovery of a New Absorption Feature at $\lambda = 2800$Å" in *Astrophysics and Space Science*, volume 94, issue number 1, pages 223-229, https://adsabs.harvard.edu/pdf/1983Ap%26SS..94..223K.

Kragh, Helge. (1993). "Big Bang Cosmology" (pages 371-390) in *Cosmology: Historical, Literary, Philosophical, Religious, and Scientific Perspectives*, edited by Norriss S. Hetherington. New York: Garland Publishing.

Kragh, Helge. (1993). "Steady State Theory" (pages 391-406) in *Cosmology: Historical, Literary, Philosophical, Religious, and Scientific Perspectives*, edited by Norriss S. Hetherington. New York: Garland Publishing.

Kragh, Helge. (1996). *Cosmology and Controversy: The Historical Development of Two Theories of the Universe*. Princeton, New Jersey: Princeton University Press.

Kragh, Helge S. (2010). "When is a Prediction Anthropic? Fred Hoyle and the 7.65 MeV Carbon Resonance" in *PhilSci-Archive – An Archive for Preprints in Philosophy of Science*: University of Pittsburgh University Library System, pages 1-36, http://philsci-archive.pitt.edu/5332/1/3alphaphil.pdf.

Kragh, Helge. (November 2010). "An Anthropic Myth: Fred Hoyle's Carbon-12 Resonance Level" in *Archives for History of Exact Sciences*, volume 64, issue number 6, pages 721-751, https://link.springer.com/article/10.1007/s00407-010-0068-8.

Kragh, Helge S. (February-March 2011). "The Origin of the Modern Anthropic Principle" in *The Journal of Cosmology*, volume 13, issue number 1, pages 3700-3705, http://cosmology.com/Anthropic100.html.

Kragh, Helge, and Robert W. Smith. (June 2003). "Who Discovered the Expanding Universe?" in *History of Science*, volume 41, part 2, issue number 132, pages 141-162, https://journals.sagepub.com/doi/abs/10.1177/007327530304100202.

Kragh, Helge S., and Dominique Lambert. (10 October 2007). "The Context of Discovery: Lemaitre and the Origin of the Primeval-Atom Universe" in *Annals of Science*, volume 64, issue number 4, pages 445-470, doi:10.1080/00033790701317692, https://www.tandfonline.com/doi/abs/10.1080/00033790701317692.

Kwok, Sun. (2008). "Synthesis of Organic Compounds in the Late Stages of Stellar Evolution and Their Connection to the Solar System" [a Conference Paper] in *Proceedings of the 10th Asian-Pacific Regional International Astronomical Union* [IAU] *Meeting* (APRIM 2008), Kunming, China, 3-6 August 2008, and in Faculty of Science: Conference Papers: HKU Scholars Hub, University of Hong Kong.

Kwok, Sun. (2009). "Delivery of Complex Organic Compounds from Planetary Nebulae to the Solar System" in *International Journal of Astrobiology*, volume 8, issue number 3, pages 161-167, doi:10.1017/S1473550409004492, https://doi.org/10.1017S1473550409004492.

Kwok, Sun. (2009). "Organic Matter in Space: From Star Dust to the Solar System" in *Astrophysics and Space Science*, volume 319, issue number 1, pages 5-21, doi:10.1007/s10509-008-9965-6, online at https://link.springer.com/article/10.1007/s10509-008-9965-6.

Kwok, Sun. (September-October 2011). "From 'Frontiers of Astronomy' to Astrobiology" in *The Journal of Cosmology*, volume 16, issue number 13, pages 6643-6660, online at https://thejournalofcosmology.com/13_Kwok.pdf.

Kwok, Sun. (December 2011 [Received 3 August 2011]). "Delivery of Complex Organic Compounds from Evolved Stars to the Solar System" in *Origins of Life and Evolution of Biospheres*, volume 41, issue number 6, pages 497-502, doi:10.1007/s11084-011-9254-1, online at https://link.springer.com/article/10.1007/s11084-011-9254-1.

Kwok, Sun. (2012). *Organic Matter in the Universe*. Weinheim, Germany: Wiley-VCH Publishing.

Kwok, Sun and Scott A. Sandford, editors. (2008). *Organic Matter in Space: Proceedings of the 251st Symposium of the International Astronomical Union Held in Hong Kong, China, February 18-22, 2008*. Cambridge, England: Cambridge University Press.

Lal, Ashwini Kumar, and Rhawn Joseph. (30 January 2010). "Big Bang? A Critical Review" in *The Journal of Cosmology*, volume 6, issue number 11, pages 1533-1547, online at https://thejournalofcosmology.com/BigBang101.html.

Lemaître, Georges Edouard. (1927). "Un Univers homogène de masse constant et de rayon croissant rendant compte de la 47arvard radiale des nébuleuses extragalactiques" [A homogeneous Universe of constant mass and growing radius accounting for the radial velocity of extragalactic nebulae]. *Annales de la Société Scientifique de Bruxelles*.

Lemaître, G. [Georges Edouard]. (9 May 1931). "The Beginning of the World from the Point of View of Quantum Theory" in *Nature*, volume 127, issue number 3210, page 706-706, https://www.nature.com/articles/127706b0.

Lemaître, Georges Edouard. (1946). *L'Hypothèse de l'Atome Primitive: Essai de Cosmogonie*. Neuchatel, Switzerland: Éditions du Griffon. [Kragh 1996: 469] [J. J. O'Connor, E. F. Robertson July 2008]

Lemaître, Georges Edouard. (1950). *The Primeval Atom: An Essay on Cosmogony* [*L'Hypothèse de l'atome primitive: Essai de Cosmogonie* (1946)], translation by Betty H. and Serge A. Korff, Preface by Ferdinand Gonseth, Foreword by Henry Norris Russell. New York: D. Van Nostrand Company.

Lemaître, Georges Edouard, and others. (1933). *Discussion sur l'évolution de l'univers*. Paris, France: Gauthier-Villars Publishing.

Lerner, Eric J. (1992 [1991]). *The Big Bang Never Happened*, paperback with new author Preface. New York: Random House Vintage Books.

Levin, Michael. (December 2019). "The Computational Boundary of a 'Self': Developmental Bioelectricity Drives Multicellularity and Scale-Free Cognition" in *Frontiers in Psychology* [DOI: 10.3389/fpsyg.2019.02688].
Also, see "What are Cognitive Light Cones? (Michael Levin Interview)" (2023) on YouTube
at https://www.youtube.com/watch?v=YnObwxJZpZc&t=19s, and see "What is The Field of Diverse Intelligence? Hacking the Spectrum of Mind & Matter | Michael Levin" (August 2023) on YouTube at
https://www.youtube.com/watch?v=kMxTS7eKkNM&t=364s.

Longair, Malcolm S. (2005). "Evolutionary Cosmologies: Then and Now" is chapter 8 in *The Scientific Legacy of Fred Hoyle*, edited by Douglas Gough. Cambridge: Cambridge University Press, 2011 paperback.

Manning, Kenneth R. (1983). *Black Apollo of Science: The Life of Ernest Everett Just*. Oxford: Oxford University Press.

Maddox, John. (1 September 1994). "The Return of Cosmological Creation" in *Nature*, volume 371, issue number 6492, page 11-11, https://ui.adsabs.harvard.edu/abs/1994Natur.371...11M/abstract.

Maddox, John. (20 September 2001). "Obituary: Fred Hoyle (1915-2001)" in *Nature*, volume 413, issue number 6853, page 270-270, doi:10.1038/35095162, https://www.nature.com/articles/35095162.

Maddox, John. (6 June 2002). "Astronomy: The Hoyle Story" in *Nature*, volume 417, issue number 6889, pages 603-605, doi:10.1038/417603a, https://www.nature.com/articles/417603a.

Mathis, John S., William Rumpl, and Kenneth H. Nordsieck. (15 October 1977). "The Size Distribution of Interstellar Grains" in *the Astrophysical Journal*, Part 1, volume 217, pages 425-433,

doi:10.1086/155591,
https://adsabs.harvard.edu/pdf/1977ApJ...217..425M.

Mautner, Michael N. (February-March 2010). "Seeding the Universe with Life: Securing Our Cosmological Future" in *The Journal of Cosmology*, volume 5, issue number 26, pages 982-994, https://thejournalofcosmology.com/SearchForLife111.html.

McConnell, Craig. (October 2006). "The BBC, the Victoria Institute, and the Theological Context for the Big Bang – Steady State Debate" in *Science and Christian Belief*, volume 18, issue number 2, pages 151-168.

McNaughton, N. J. and C. T. Pillinger. (11 December 1980). "Comets and the Origin of Life" in *Nature*, volume 288, issue number 5791, page 540-540, doi:10.1038/288540a0, online at https://www.nature.com/articles/288540a0.

Miller, Stanley L. and Harold C. Urey. (July 1959). "Organic Compound Synthesis on the Primitive Earth: Several Questions about the Origin of Life Have Been Answered, But Much Remains to be Studied" in *Science*, volume 130, issue number 3370, pages 245-251, online at https://www.science.org/doi/abs/10.1126/science.130.3370.245.

Miller, Stanley L. and Leslie E. Orgel. (1974). *The Origins of Life on the Earth*. Upper Saddle River, New Jersey: Prentice Hall Publisher.

Mitton, Simon. (2005). *Conflict in the Cosmos: Fred Hoyle's Life in Science*. Washington, DC: Joseph Henry Press.

Mitton, Simon. (2005). *Fred Hoyle: A Life in Science*. London: Aurum Press [reprint, Cambridge University Press, 2011].

Mitton, Simon. (2008). "Hoyle, Fred" (pages 388-392) in *New Dictionary of Scientific Biography*. Detroit: Charles Scribner's Sons.

Napier, W. M. (February 2004 [Online 30 January 2004]). "A Mechanism for Interstellar Panspermia" in *Monthly Notices of the Royal Astronomical Society,* volume 348, issue number 1, pages 46-51, https://academic.oup.com/mnras/article/348/1/46/1415892.

Napier, W. M., J. T. Wickramasinghe, and N. C. Wickramasinghe. (2004 [Received 12 August 2004]). "Extreme Albedo Comets and the Impact Hazard" in *Monthly Notices of the Royal Astronomical Society*, volume 355, issue number 1, pages 191-195, doi:10.1111/j.1365-2966.2004.08309.x, https://academic.oup.com/mnras/article/355/1/191/3101517.

Narlikar, Jayant V. (1973). "Steady State Defended" (pages 69-84) in *Cosmology Now*, edited by L. John. London: BBC. [Kragh 1996: 467, 474]

Narlikar, Jayant V. (2005). "Alternative Ideas in Cosmology" is chapter 9 in *The Scientific Legacy of Fred Hoyle*, edited by Douglas Gough. Cambridge: Cambridge University Press, 2011 paperback.

Narlikar, Jayant V., and Nalin Chandra Wickramasinghe. (7 October 1967). "Microwave Background in a Steady State" in *Nature*, volume 216, issue number 5110, pages 43-44, doi:10.1038/216043a0, online at https://www.nature.com/articles/216043a0.

Narlikar, Jayant V., and N. Chandra Wickramasinghe. (30 March 1968). "Interpretation of Cosmic Microwave Background" in *Nature*, volume 217, issue number 5135, pages 1235-1236, doi:10.1038/2171235a0, https://www.nature.com/articles/2171235a0.

Narlikar, Jayant V. (September 1992). "The Concepts of 'Beginning' and 'Creation' in Cosmology" (pages 361-371) in *Philosophy of Science*, volume 59, issue number 3, pages 361-371.

Narlikar, Jayant V., Indu Banga, and Chanda Gupta, editors. (1992). *Philosophy of Science: Perspectives from Natural and Social Sciences*. Shimla, India: Indian Institute of Advanced Study; Delhi, India: Munshiram Manoharlal Publishers.

Narlikar, Jayant V., and Geoffrey Burbidge. (2008). "An Alternative Cosmology" is chapter 15 in *Facts and Speculations in Cosmology*. Cambridge: Cambridge University Press.

O'Connor, J. J. and E. F. Robertson. (July 2008). "Georges Henri-Joseph-Edouard Lemaître" in *MacTutor History of Mathematics* online at www-history.mcs.st-andrews.ac.uk/Biographies/Lemaitre.html.

Ogden, Schubert M. (Spring 1984). "Process Theology and the Wesleyan Witness" in *Perkins School of Theology Journal*, volume 37, number 3, pages 18-33 Reprinted with a collection of essays responsive to Ogden's essay in *Thy Nature and Thy Name Is Love: Wesleyan and Process Theologies in Dialogue*, (Nashville: Abingdon Press, 2001), edited by Bryan P. Stone and Thomas Jay Oord.

Oparin, Alexander Ivanovitch. (1924). *The Origin of Life*. Moscow: Moscow Worker publisher, 1924 (in Russian); English translation: New York: Dover (1952 [1938]).
Also, see "ALEXANDER OPARIN (1894-1980)" online at https://www.physicsoftheuniverse.com/scientists_oparin.html, where is said: "In 1924, Oparin officially put forward his influential theory that life on Earth developed through gradual chemical evolution of carbon-based molecules in a "primordial soup", at just about the same time as the British biologist J. B. S. Haldane was independently proposing a similar theory." (Accessed 11 January 2023)]

Oparin, Aleksandr Ivanovičh. (1938 [1936]). *The Origin of Life*, translation with annotations by Sergius Morgulis. New York: Macmillan Publishers; Reprint, Mineola, New York: Courier Dover Publishers, 1953 and 2003.

Oparin, Aleksandr Ivanovičh. (1968 [1966]). *Genesis and Evolutionary Development of Life*. New York: Academic Press, 2017.

Pasteur, Louis. (6 February 1860). "Expériences relatives aux générations spontanées" in *Comptes rendus de l'Académie des Sciences* [Proceedings of the [French] Academy of Sciences], volume 50, pages 303-307.

Pasteur, Louis. (7 May 1860). "De l'origines des ferments: Nouvelle experiences relatives aux generations dites spontanées" in *Comptes rendus de l'Académie des Sciences* - Proceedings of the [French] Academy of Sciences, 3, volume 50, pages 849-854.

Pasteur, Louis. (3 September 1860). "Nouvelle experiences relatives aux generations dites spontanées" in *Comptes rendus de l'Académie des Sciences* [Proceedings of the [French] Academy of Sciences], volume 51, pages 348-353.

Pasteur, Louis. (5 November 1860). "Suite à uneprécédente communication relative aux generations dites spontanées" in *Comptes rendus de l'Académie des Sciences* [Proceedings of the [French] Academy of Sciences], volume 51, pages 675-678.

Pasteur, Louis. (1861). "Mémoiresur les corpuscules organisés qui existent dans l'atmosphère: Examen de la doctrine des generations spontanées" in *Annales des Sciences Naturelles* (partie Zoologique), 4e série, volume 16, pages 5-68.

Penrose, Roger. (1991). *The Emperor's New Mind: Concerning Computers, Minds, and the Laws of Physics*, Foreword by Martin Gardner. New York: Penguin Books [originally Oxford University Press, 1989].

Penrose, Roger. (1994). *Shadows of the Mind: A Search for the Missing Science of Consciousness*. Oxford: Oxford University Press.

Penrose, Roger. (2004). "Speculative Theories of the Early Universe" and "The Anthropic Principle" in *The Road to Reality: A Complete Guide to the Laws of the Universe*. London: Jonathan Cape Publisher.

Penrose, Roger. (2010). "Conformal Cyclic Cosmology" in *Cycles of Time: An Extraordinary New View of the Universe*. London: Bodley Head Publisher; Reprint, New York: Alfred A. Knopf, 2011.

Penrose, Roger. (2011). *Collected Works*. Oxford; New York: Oxford University Press.

Penrose, Roger, and Wolfgang Rindler. (1984, 1986). *Spinors and Space-time*, volumes 1 and 2. Cambridge; New York: Cambridge University Press.

Penrose, Roger, and others. (1999 [originally 1997]). *The Large, the Small, and the Human Mind*, new and revised, edited by Malcolm Longair. Cambridge; New York: Cambridge University Press.

Pflug, Hans Dieter. (26 November 1981). "Extraterrestrial Life: New Evidence of Microfossils in the Murchison Meteorite." A public lecture in Cardiff.

Pflug, Hans Dieter. (1984). "Ultrafine Structure of the Organic Matter in Meteorites" in *Fundamental Studies and the Future of Science*, edited by Chandra Wickramasinghe. Cardiff, Wales, United Kingdom: University College Cardiff Press.

Pflug, Hans Dieter and H. Haescheke-Boyer. (9 August 1979). "Combined Structural and Chemical Analysis of 3,800-Myr-Old Microfossils" in *Nature*, volume 280, issue number 5722, pages 483-486, doi:10.1038/280483a0, https://www.nature.com/articles/280483a0.

Polkinghorne, John C. (1989). *Science and Providence: God's Interaction with the World*. Boston: New Science Library.
For a review of *Science and Providence*, see "Articles of Faith" (4 May 1989) by Fred Hoyle in *Nature*, volume 339, issue number 6219, pages 23-24.

Ponnamperuma, Cyril, editor. (c1981). *Comets and the Origin of Life: Proceedings of the Fifth College Park Colloquium on Chemical Evolution, University of Maryland, College Park, Maryland, U.S.A., October 29th to 31st, 1980*. Boston: Kluwer Academic Publishers.

Powell, Russell. (2020). *Contingency and Convergence: Toward a Cosmic Biology of Body and Mind*. Cambridge, Massachusetts: The MIT Press.

Raymo, Chat. (February 2005). "Big Bang vs. Steady State: How the Big Bang Theory Won the 20th Century's Biggest Cosmological Debate" [book review of *Big Bang: The Origin of the Universe* (HarperCollins 2005) by Simon Singh] in *Scientific American*, volume 292, issue number 2, pages 98-100, doi:10.1038/scientificamerican0205-98.

Rees, Martin. (1997). *Before the Beginning: Our Universe and Others*, Foreword by Stephen Hawking. Cambridge, Massachusetts: Helix Books.

Rees, Martin. (2001). *Our Cosmic Habitat*. Princeton, New Jersey: Princeton University Press.

Rees, Martin (2005). Foreword (pages x-xiii) in *The Scientific Legacy of Fred Hoyle*, edited by Douglas Gough. Cambridge: Cambridge University Press, 2011 paperback.

Rees, Martin. (10 January 2011). "Life in the Cosmos" is a Madingley Lecture at University of Cambridge.

Rees, M. J., and J. P. Ostriker. (June 1977 [Received 5 November 1976 [original form 7 July 1976]). "Cooling, Dynamics and Fragmentation of Massive Gas Clouds: Clues to the Masses and Radii of Galaxies and Clusters" in *Monthly Notices of the Royal Astronomical Society*, volume 179, pages 541-559.

Richards, Robert J. (1987). *Darwin and the Emergence of Evolutionary Theories of Mind and Behavior*. Chicago and London: University of Chicago Press.

Rowan-Robinson, Michael. (22 December 1972). "Steady State Obituary?" in *Nature*, volume 240, issue number 5382, page 439-439, online at https://www.nature.com/articles/240439a0.

Rushton, Simon K. and Rob Gray. (5 October 2006). "Hoyle's Observations Were Right on the Ball" in *Nature*, volume 443, issue number 7111, page 506-506, doi:10.1038/443506d, online at https://www.nature.com/articles/443506d.

Ryle, Martin. (August 1955). "Radio Stars and Their Cosmological Significance" in *The Observatory*, volume 75, issue number 887, pages 137-147, online at https://adsabs.harvard.edu/pdf/1955Obs....75..137R.

Ryle, Martin. (3 June 1961). "Radio Astronomy and Cosmology" in *Nature*, volume 190, issue number 4779, pages 852-854, doi:10.1038/190852a0, online at https://link.springer.com/article/10.1038/190852a0.

Ryle, Martin, and R. W. Clarke (30 January 1961). "An Examination of the Steady-State Model in the Light of Some Recent Observations of Radio

Sources" in *Monthly Notices of the Royal Astronomical Society*, volume 122, issue number 4, pages 349-362, https://academic.oup.com/mnras/article/122/4/349/2602359.

Sagan, Carl, and Ann Druyan. (1985). *Comet*. New York: Random House Publishing [reprint, Ballantine Books, 1997].

Saint John's College University of Cambridge "Preprints of Sir Fred Hoyle (1915-2001), astronomer" from the Astrophysics and Relativity Preprint Series and Special Preprints, produced by the Department of Applied Mathematics and Astronomy, University College, Cardiff and the Mathematics Institute, Cardiff, https://www.joh.cam.ac.uk/library/special_collections/personal_papers/hoylepreprints.

Sakata, A., N. Nakagawa, T. Iguchi, S. Isobe, M. Morimoto, F. Hoyle, and N. C. Wickramasinghe. (17 March 1977). "Spectroscopic Evidence for Interstellar Grain Clumps in Meteoritic Inclusions" [Letter to Nature] in *Nature*, volume 266, issue number 5599, page 241-241, https://www.nature.com/articles/266241a0.

Salpeter, E. E. and N. C. Wickramasinghe. (3 May 1969). "Alignment of Interstellar Grains by Cosmic Rays" in *Nature*, volume 222, issue number 5192, pages 442-444, doi:10.1038/222442a0, https://www.nature.com/articles/222442a0.

Sandford, Scott A. (2008). "Organics in the Samples Returned from Comet 81P/Wild 2 by the Stardust Spacecraft" (pages 299-308) in *Organic Matter in Space: Proceedings of the 251st Symposium of the International Astronomical Union Held in Hong Kong, China, February 18-22, 2008*, edited by Sun Kwok and Scott A. Sandford. Cambridge: Cambridge University Press.

Sargent, Wallace L. W. (2005). "Fred Hoyle's Major Work in the Context of Astronomy and Astrophysics Today" is chapter 1 in *The Scientific Legacy of Fred Hoyle*, edited by Douglas Gough. Cambridge: Cambridge University Press, 2011 paperback.

Savage, Blair. D., and John S. Mathis. (1979). "Observed Properties of Interstellar Dust" in *Annual Review of Astronomy and Astrophysics*, volume 17, pages 73-111, doi:10.1146/annurev.aa.17.090179.000445, https://adsabs.harvard.edu/pdf/1979ARA%26A..17...73S.

Scalzi, Giuliano, Laura Selbmann, Laura Zucconi, Elke Rabbow, Gerda Horneck, Patrizia Albertano, and Silvano Onofri. (12 June 2012 [Received 5 December 2011]). "LIFE Experiment: Isolation of Cryptoendolitic Organisms from Antarctic Colonized Sandstone Exposed to Space and Simulated Mars Conditions on the International Space Station" in *Origins of Life and Evolution of Biospheres*, volume 42, issue number 2, pages 253-262, doi:10.1007/s11084-012-9282-5, https://link.springer.com/article/10.1007/s11084-012-9282-5.

Schalén, C. (1939). *Uppsala Obs*. Ann. 1, No. 2.
"In Astronomy, 1939 was the year when a Swedish astronomer by name C. Schalén first showed that the universe contained vast amounts of cosmic dust (microscopic dust particles) that blocked out the light of distant stars ..." (Wickramasinghe, December 2011).

Schild, Rudolf E. (June 2012). "COMMENTARY: Introduction to Astro-Theology" in *The Journal of Cosmology*, volume 19, issue number 1, pages 8547-8551,
https://thejournalofcosmology.com/indexVol19CONTENTS.htm,
https://thejournalofcosmology.com/commentary.pdf.

Schild, Rudolf, R. Gabriel Joseph, Chandra Wickramasinghe, Robert J. C. McLean, Richard B. Hoover, Gilbert V. Levin, Milton Wainwright, and Max K. Wallis. (2014 [2010]). *Biological Cosmology, Astrobiology, Extraterrestrial Life*. Cosmology Science Publishers.

Shapiro, Robert (13 May 1993). "Life, the Universe and Anything Goes" [a review of *Our Place in the Cosmos* (1993) by Fred Hoyle and N. Chandra Wickramasinghe] in *Nature*, volume 363, issue number 6425, page 124-124, doi:10.1038/363124a0,
https://www.nature.com/articles/363124a0.

Shaviv, Giora. (2012). *The Synthesis of the Elements: The Astrophysical Quest for Nucleosynthesis and What It Can Tell Us About the Universe*. Heidelberg, Germany: Springer-Verlag Publisher.

Shivaji, Sisinthy. Sreenivas Ara, Snajay Kumar Singh, Sunil Bandi, Aditya Singh, and Anil Kumar Pinnaka. (December 2012 [9 November 2012]). "Draft Genome Sequence of Bacillus Isronensis Strain B3W22, Isolated from the Upper Atmosphere" in *Journal of Bacteriology*, volume 194, issue number 23, pages 6624-6625, doi:10.1128/JB.01651-12, https://journals.asm.org/doi/full/10.1128/JB.01651-12.

Shivaji, Sisinthy with Preeti Chaturvedi, Zareena Begum, Pavan Kumar Pindi, R. Manorama, D. Ananth Padmanaban, Yogesh S. Shouche, Shrikant Pawar, Parag Vaishampayan, C. B. S. Dutt, G. N. Datta, R. K. Manchanda, U. R. Rao, P. M. Bhargava and J. V. Narlikar. (2009). "*Janibacter hoylei sp. nov.* [a new species of bacteria named for Fred Holye], *Bacillus isronensis sp. nov.* and *Bacillus aryabhattai sp. nov.*, isolated from cryotubes used for collecting air from the upper atmosphere" in the *International Journal of Systematic and Evolutionary Microbiology*, volume 59, issue number 12, pages 2977-2986, doi:10.1099/ijs.0.002527-0, https://www.microbiologyresearch.org/content/journal/ijsem/10.1099/ijs.0.002527-0.

Singh, Simon. (2005). *Big Bang: The Origin of the Universe*. New York: HarperCollins.

Slate, Nico. (2014). *The Prism of Race: W.E.B. Du Bois, Langston Hughes, Paul Robeson, and the Colored World of Cedric Dover*. New York: Palgrave Macmillan.

Smith, Eric, and Harold J. Morowitz. (2022 [first edition 2016]). *The Origin and Nature of Life on Earth: The Emergence of the Fourth Geosphere*. Cambridge, UK: Cambridge University Press.
Smith and Morowitz argue that "the emergence of life was a necessary cascade of non-equilibrium phase transitions that opened new channels for chemical energy flow on Earth" (2022: i) and that "the core of intermediary metabolism is *a necessary consequence of galactic processes*" (2022: xvi). [Italics added.] Though not mentioned by Smith and Morowitz, their arguments are consistent with the generic idea of panspermia (seeds of potential new life panoramically distributed) and

the Hoyle-Wickramasinghe conviction that life is a stellar-galactic and cosmic phenomenon.

Smolin, Lee. (29 July 2004). "Scientific Alternatives to the Anthropic Principle" in *Universe or Multiverse* (21 June 2007), edited by Bernard Carr. Cambridge: Cambridge University Press, pages 323-366.

Steele, Edward J., Shirwan Al-Mufti, Kenneth A. Augustyn, Rohana Chandrajith, John P. Coghlan, S. G. Coulson, Sudipto Ghosh, Mark Gillman, Reginald M. Gorczynski, Brig Klyce, Godfrey Louis, Kithsir Mahanama, Keith R Oliver, Julio Padron, Jiangwen Qu, John A Schuster, W. E. Smith, Duane P Snyder, Julian A. Steele, Brent J. Stewart, Robert Temple, Gensuke Tokoro, Christopher A Tout, Alexander Unzicker, Milton Wainwright, Jamie Wallis, Daryl H. Wallis, Max K. Wallis, John Wetherall, D. T. Wickramasinghe, J. T. Wickramasinghe, N. C. Wickramasinghe, and Yongsheng Liu. (August 2018). "Cause of Cambrian Explosion - Terrestrial or Cosmic?" in *Progress in Biophysics and Molecular Biology*, volume 136, pages 3-23, https://www.sciencedirect.com/science/article/pii/S0079610718300798.

Steele, Edward J., Reginald M Gorczynski, Robyn A Lindley Yongsheng Liu, Robert Temple, Gensuke Tokoro, Dayal T Wickramasinghe, and N Chandra Wickramasinghe.
(December 2019). "Lamarck and Panspermia – On the Efficient Spread of Living Systems throughout the Cosmos" in *Progress in Biophysics & Molecular Biology* - An International Review Journal, volume 149, pages 10-32, doi:10.1016/j.pbiomolbio.2019.08.010, https://www.sciencedirect.com/science/article/pii/S0079610719301129.

Tepfer, David, and Sydney Leach. (December 2006 [Received 22 June 2006, accepted 16 August 2006, published 15 November 2006]).
"Plant Seeds as Model Vectors for the Transfer of Life Through Space" in *Astrophysics and Space Science*, volume 306, issue number 1, pages 69-75, doi:10.1007/s10509-006-9239-0, https://link.springer.com/article/10.1007/s10509-006-9239-0.

Thomas, Paul J. with Christopher F. Chyba, and Christopher P. McKay, editors. (1997). *Comets and the Origin and Evolution of Life*. New York: Springer [Reprint, Berlin: Springer-Verlag, 2006].

Tirard, Stéphane. (24 November 2017). "J. B. S. Haldane and the origin of life" in *Journal of Genetics*, volume 96, pages 735-739, online at https://pubmed.ncbi.nlm.nih.gov/29237880/.
Abstract - In 1929 the British biologist John Burdon Sanderson Haldane published a hypothesis on the origin of life on earth, which was one of the most emblematic of the interwar period. It was a scenario describing the progressive evolution of matter on the primitive earth and the emergence of life. Firstly, this paper presents the main ideas put forward by Haldane in this famous text. The second part makes comparisons between Haldane and Alexander Ivanovitch Oparin's ideas regarding the origins of life (1924). These two theories, apparently very similar, presented distinct conclusions. The third part focusses on Haldane's reflections on the emergence of life during the 1950s and 1960s, and shows how they were linked to the recent developments of prebiotic chemistry and molecular biology.

Tyson, Neil de Grasse, and Donald Goldsmith. (2004). *Origins: Fourteen Billion Years of Cosmic Evolution*. New York; London: W. W. Norton.

Urey, Harold Clayton. (1952). *The Planets, Their Origin and Development*. New Haven, Connecticut: Yale University Press.

Urey, Harold C. (24 March 1962). "Life-Forms in Meteorites: Origin of Life-like Forms in Carbonaceous Chondrites Introduction" in *Nature*, volume 193, issue number 4821, pages 1119-1123, doi:10.1038/1931119a0, https://www.nature.com/articles/1931119a0.

Urey, Harold Clayton (1963). *Some Cosmochemical Problems*. University Park, Pennsylvania: Pennsylvania State University.

Vanysek, V. and N. C. Wickramasinghe. (October 1999 [originally April 1975]). "Formaldehyde Polymers in Comets" [Letter to the Editor] in *Astrophysics and Space Science*, volume 268, issue number 1, pages 115-124 [originally, volume 33, issue number 2, L19-L28], doi:10.1023/A:1002452904168, https://link.springer.com/article/10.1023/A:1002452904168.

Vsekhsvyatskiy, Sergei Konstantinovich. (1970). *The Nature and Origin of Comets and Meteors*. Washington, DC: National Aeronautics and Space Administration.

Wagoner, Robert V., William A. Fowler, and Fred Hoyle. (April 1967 [Received 1 September 1966]). "On the Synthesis of Elements at Very High Temperatures" in *the Astrophysical Journal*, volume 148, issue number 1, pages 3-49, doi:10.1086/149126, https://adsabs.harvard.edu/pdf/1967ApJ...148....3W.

Wainwright, Milton, Fawaz Alshammari, and Khalid Alabri. (May 2010). "Are Microbes Currently Arriving to Earth from Space?" in *The Journal of Cosmology*, volume 7, issue number 3, pages 1692-1702, https://thejournalofcosmology.com/Panspermia2.html.

Wainwright, Milton, Christopher E. Rose, Alexander J. Baker, and N. Chandra Wickramasinghe. (January 2014). "Impact Events on a Graphite Stub Provide Evidence That a Biological Entity Arrived at the Stratosphere from Space" in *The Journal of Cosmology*, volume 23, issue number 7, pages 11131-11135, online at https://thejournalofcosmology.com/joc-32_wainwright.pdf.

Wainwright, Milton, N. C. Wickramasinghe, J. V. Narlikar, and P. Rajaratnam. (January 2003 [Received 24 September 2002, revised 31 October 2002, accepted 5 November 2002, online 3 December 2002]). "Microorganisms Cultured from Stratospheric Air Samples Obtained at 41km" in *Federation of European Microbiological Societies - Microbiology Letters*, volume 218, issue number 1, pages 161-165, https://academic.oup.com/femsle/article/218/1/161/532689.

Wainwright, Milton, and N. Chandra Wickramasinghe, Foreword by Gensuke Tokoro. (2023). *Life Comes from Space: The Decisive Evidence*. New Jersey, London: World Scientific.

Walker, Theodore, Jr. (June 2012). "The Liberating Role of Astronomy in an Old Farmer's Almanac: David Rittenhouse's 'Useful Knowledge' and a Benjamin Banneker Almanac for 1792" in *The Journal of Cosmology*, volume 19, online at https://thejournalofcosmology.com/walker4a.cor3.pdf.

Abstract - Traditionally, astronomy met theology and political ethics in almanacs. As presented in early New England almanacs of the farmer's type, astronomy was deity-affirming and liberty-oriented. The old English label for astronomy that affirms theology was "Astro-theology" (William Derham, 1715). The New England rendering of astro-theology was so strongly oriented towards liberty that it can now be labeled astro-liberation theology. This 21st century label is appropriate because 18th century New England printers and astronomers used astronomy to demonstrate the glory of the Creator (astro-theology) and to encourage liberation from colonialism and slavery (astro-liberation theology). A philosophy of astronomy as "useful knowledge" expressed by David Rittenhouse in 1775—and implicit in a Benjamin Banneker almanac for 1792—included liberty-oriented visions of planet Earth as seen from outer space, and liberty-oriented visions of intelligent life on other planets orbiting other stars.

Walker, Theodore, Jr. (29 April 2020). "Interdisciplinary Convergences with Biology and Ethics via Cell Biologist Ernest Everett Just and Astrobiologist Sir Fred Hoyle" (chapter 2, pages 11-35) in *Panentheism and Panpsychism: Philosophy of Religion Meets Philosophy of Mind*; Series: Innsbruck Studies in Philosophy of Religion, Volume: 2. Brill | mentis, 2020), edited by Godehard Brüntrup, Benedikt Paul Göcke, and Ludwig Jaskolla.

Walker, Theodore, Jr. (December 2021). "Reviewing Ernest Everett Just's *Biology of the Cell Surface* (1939) and related literature, plus annotated references, hereby advancing evolutionary biology and evolutionary bioethics" in *SCIREA Journal of Health*, volume 5, issue number 6, pages 123-144, https://www.scirea.org/journal/PaperInformation?PaperID=6569.

Walker, Theodore, Jr. (27 June 2023). "Evolutionary Biology with Prophetic Structure and Content" [review article about *A Voice in the Wilderness: A Pioneering Biologist Explains How Evolution Can Help Us Solve Our Biggest Problems* (2022) by Joseph L. Graves Jr.] in *Black Theology: An International Journal* doi: 10.1080/14769948.2023.2228623, online at https://doi.org/10.1080/14769948.2023.2228623.

Walker, Theodore, Jr., and Chandra Wickramasinghe, with editing by Alexander Vishio. (2015). *The Big Bang and God: An Astro-Theology wherein an astronomer and a theologian offer a study of interdisciplinary convergences with natural theology both in the scientific researches of Sir Fred Hoyle and in the philosophical researches of Charles Hartshorne and Alfred North Whitehead, thereby illustrating a constructive postmodern trend.* New York: Palgrave Macmillan.

Wallis, Jamie with Nori Miyake, Richard B. Hoover, Andrew Oldroyd, Daryl H. Wallis, Anil Samaranayake, K. Wickramarathne, M. K. Wallis, Carl H. Gibson and Nalin Chandra Wickramasinghe. (5 March 2013). "The Polonnaruwa Meteorite: Oxygen Isotope, Crystalline and Biological Composition" in *The Journal of Cosmology*, volume 22, issue number 2, pages 10004-10011, https://thejournalofcosmology.com/Paper22(2a).pdf.

Wallis, Max K. (3 April 1980). "Radiogenic Melting of Primordial Comet Interiors" in *Nature*, volume 284, issue number 5755, pages 431-433, doi:10.1038/284431a0, https://www.nature.com/articles/284431a0.

Wallis, Max K. (17 July 1980). "Cometary Science" in *Nature*, volume 286, issue number 5770, pages 207-208, doi:10.1038/286207a0, https://www.nature.com/articles/286207a0.

Wallis, Max K., and N. C. Wickramasinghe, F. Hoyle, and R. Rabilizirov. (January 1989 [Received 29 December 1988]). "Biologic Verses Abiotic Models of Cometary Grains" in *Monthly Notices of the Royal Astronomical Society*, volume 238, issue number 4, pages 1165-1170, https://academic.oup.com/mnras/article/238/4/1165/1037496.

Wallis, Max K., and N. C. Wickramasinghe. (April 1991 [Received 17 July 1990]). "Structural Evolution of Cometary Surfaces" in *Space Science Reviews*, volume 56, issue number 1, pages 93-97, doi:10.1007/BF00178395, https://link.springer.com/article/10.1007/BF00178395.

Wallis, Max K., N. C. Wickramasinghe, and F. Hoyle. (1992). "Cometary Habitats for Primitive Life" in *Advances in Space Research*, volume 12, issue number 4, pages 281-285, https://www.sciencedirect.com/science/article/abs/pii/027311779290184Y.

Wallis, Max K., and N. C. Wickramasinghe. (1995). "Role of Major Terrestrial Cratering Events in Dispersing Life in the Solar System" in *Earth and Planetary Science Letters*, volume 130, pages 69-73, https://www.sciencedirect.com/science/article/abs/pii/0012821X9400232N.

Wallis, Max K. and Sirwan Al-Mufti. (February 1996). "Processing of Cometary Grains at the Nucleus Surface" in *Earth, Moon, and Planets*, volume 72, issue number 1, pages 91-97, doi:10.1007/BF00117507, https://link.springer.com/article/10.1007/BF00117507.

Wallis, Max K., and Nalin Chandra Wickramasinghe. (2004). "Interstellar Transfer of Planetary Microbiota" in *Monthly Notices of the Royal Astronomical Society*, volume 348, issue number 1, pages 52-57, 61, https://academic.oup.com/mnras/article/348/1/52/1415928.

Wassserburg, G. J. with William A. Fowler and Fred Hoyle. (February 1960 [Received 31 January 1960]). "Duration of Nucleosynthesis" in *Physical Review Letters*, volume 4, issue number 3, pages 112-114, doi: 10.1103/PhysRevLett.4.112.

Weinberg, Steven. (January 1989). "The Cosmological Constant Problem" in *Reviews of Modern Physics*, volume 61, issue number 1, pages 1-23, doi:10.1103/RevModPhys.61.1, https://journals.aps.org/rmp/abstract/10.1103/RevModPhys.61.1.

Weinberg, Steven. (1993 [first edition 1977]). *The First Three Minutes: A Modern View of the Origin of the Universe*, updated edition. New York: Basic Books.

Weinberg, Steven. (October 1994). "Life in the Universe" in *Scientific American*, volume 271, issue number 4, pages 44-49, https://www.jstor.org/stable/24942868.

Weinberg, Steven. (2008). *Cosmology*. Oxford: Oxford University Press.

White, Simon D. M., and Martin J. Rees. (May 1978 [Received 26 September 1977]). "Core Condensation in Heavy Halos: A Two-Stage Theory for Galaxy Formation and Clustering" in *Monthly Notices of the Royal Astronomical Society*, volume 183, issue number 3, pages 341-358, https://academic.oup.com/mnras/article/183/3/341/972568.

Whitehead, Alfred North. (1927-28). *Process and Reality: An Essay in Cosmology* (Gifford Lectures Delivered in the University of Edinburgh During the Session 1927-28), *1978 Corrected Edition*, edited by David Ray Griffin and Donald W. Sherburne. New York: Free Press. See "PART V FINAL INTERPRETATION" from *Process and Reality: An Essay in Cosmology* (1978 [1927-28]) by Alfred North Whitehead in *The Journal of Cosmology*, volume 20, online at https://thejournalofcosmology.com/Whitehead_PR_Part5_Final_Interpratation.pdf.

Whittet, D. C. B. (25 October 1979). "Interstellar Grains: Organic or Refractory?" in *Nature*, volume 281, issue number 5733, page 708-708, doi:10.1038/281708a0, online at https://link.springer.com/content/pdf/10.1038/281708a0.pdf.

Witham, Larry. (2005). *The Measure of God: Our Century-Long Struggle to Reconcile Science and Religion*. New York: HarperCollins Publishers.

Wolfram, Stephen. (2002). *A New Kind of Science*. Champaign, Illinois: Wolfram Media.

Woosley, S. E. (1999). "Hoyle and Fowler's Nucleosynthesis in Supernovae" in *Astrophysical Journal*, Centennial Issue number, volume 525, pages 924-925, https://adsabs.harvard.edu/pdf/1999ApJ...525C.924W.

Woosley, S. E. (December 2007). "Nuclear Astrophysics: The First 50 Years" in *Nature Physics*, volume 3, issue number 12, pages 832-833, doi:10.1038/nphys804, https://www.nature.com/articles/nphys804.

Wu, Katherine, and Shirlee Wohl. (4 November 2015). "Growing Together: How Viruses Have Shaped Human Evolution" from Fall 2015 Seminar Series, Science in the News – Boston (SITNBoston) on YouTube at https://www.youtube.com/watch?v=BdsRKdeA-ss&t=2297s [accessed 10 January 2023].

Zenil, Hector, editor. (2012 [Alan Turing Year.]). *A Computable Universe: Understanding and Exploring Nature as Computation*, Foreword by Roger Penrose. Hackensack, New Jersey: World Scientific Publishing.

[][][][]

6 September 2023

Milton Keynes UK
Ingram Content Group UK Ltd.
UKHW022340171124
451242UK00002B/5